TABLE OF CONTENTS

1. DATA SUMMARY

2. TOTAL DAIRY FACILITY
2.1 Herd Makeup
2.1 Feeds and Cropland
2.1 Housing
2.1 Farmstead Planning
    Site Selection
    Remodeling
    Plans, Specifications, & Contracts
    Consulting Engineers

3. REPLACEMENT ANIMAL HOUSING
3.1 Management
3.1 Space Requirements
3.1 Environment Alternatives
3.1 Housing Types
    Bedded pack
    Free stalls (solid floor)
    Totally slotted floor
    Counter-slope floor
3.2 Calf Housing
    Sanitation
    Feeding
3.3 Heifer Housing
    Feeding
3.4 Bred Heifers and Dry Cows

4. MILKING HERD FACILITIES
4.1 Stall Barns
    Stanchion Stalls
    Tie Stalls
    Stall Size
    Stall Mats
4.3 Stall Arrangement
    Barn Dimensions
    Barn Remodeling
    Stall Barn Plan
4.4 Free Stall Barns
    Free Stall Design
    Barn Design
    Barn Layout
    Barn Plan Descriptions

5. MILKING CENTER
5.1 Milking Parlors
    Parlor Types
        Flat barn
        Side-opening
        Herringbone
        Polygon
        Trigon
    Parlor Layout
    Parlor Mechanization
5.2 Holding Area
5.4 Utility Room
5.4 Storage Room
5.4 Milk Room
5.4 Toilet and Office
5.4 Milking Equipment
    Milk Cooling
    Milk Heat Recovery
5.5 Water Supply

5.6 Milking Center Construction
    Wall Construction
    Inside Lining Material
    Insulation
    Floors
    Drains
    Lighting
    Stray Voltage
5.9 Milking Center Environment
    Holding Area
    Treatment-Hospital Area
    Milking Parlor
    Milk Room
    Office
    Storage Room
    Utility Room

6. SPECIAL HANDLING AND TREATMENT
6.1 Separation
6.1 Treatment
6.1 Gang Stanchions
6.2 Maternity Area
6.2 Loading Chutes

7. BUILDING ENVIRONMENT
7.1 Housing Types
7.1 Natural Ventilating Systems
    Principles of Natural Ventilation
        Winter ventilation
        Summer ventilation
    Design
        Site selection
        Building orientation
        Ridge openings
        Eave openings
        Sidewall openings
        End walls
        Roof slope
    Management
        Winter
        Summer
7.4 Mechanical Ventilating Systems
    Design Principles
        Ventilating rates
        Temperature control
        Moisture control
    Ventilating System Types
        Exhaust ventilation
        Positive pressure ventilation
        Negative pressure with recirculation
7.7 Exhaust Ventilating System Design
    Air Inlets
        Inlet location
        Inlet size
        Inlet control
    Fans
        Fan selection
        Fan location
        Fan controls
        Fan sizes and thermostat settings
7.10 System Maintenance
7.10 Emergency Ventilation

7.10 Manure Pit Ventilation
     Annexes
     Pit Fan Installation
7.11 Attic Ventilation
7.11 Heaters
     Unit Space Heaters
     Air Make-Up Heaters
     Solar Heat
     Reusing Exhaust Air
     Heat Exchangers
7.11 Insulation
     Insulation Types
     Selecting Insulation
     Insulation Levels
     Installing Insulation
     Maximizing Insulation Effectiveness
         Vapor barriers
         Fire protection
         Doors and windows
         Birds and rodents

8. MANURE MANAGEMENT
8.1 Waste Volumes
     Milking Center Effluent
         Septic tank with absorption field
         Settling tank with surface disposal
         Lagoons
         Other methods
     Manure
     Bedding
8.2 Collection
     Slotted Floors
     Scrapers and Blades
     Front-End Loaders
     Flushing Systems
     Gravity Flow Channel
     Gravity Transfer To Storage
8.4 Storage
     Liquid Manure
     Below-Ground Storage
     Earth Storage Basins
     Above-Ground Storage
         Filling
         Agitation
         Safety
     Semi-Solid Manure
         Drained storages
         Roofed storage
     Solid Manure
8.9 Handling Manure
     Solid Manure
     Liquid Manure
         Liquids—up to 4% solids
         Slurries—4% to 15% solids
     Irrigation
8.10 Handling Lot Runoff
     Collecting Runoff
     Liquid-Solid Separation
     Holding Pond
     Vegetative Infiltration Area

8.12 Pumping Manure
     Pump Selection
         Centrifugal pumps
         Screw pumps
         Piston pumps
         Air driven pumps
         Agitation pumps
         Irrigation pumps
         Flushing and recirculating pumps
         Sumps and sump pumps
8.15 Land Application and Utilization

9. FEEDING FACILITIES
9.1 Stall Barn Feeding
9.1 Feeding With Loose Housing
9.1 Mixing and Dispensing
     Complete Mixed Rations
     Supplemental Concentrate Feeders
     Batch Mixing
     Continuous Mixing
9.3 Silage Storages
9.3 Feed Centers
9.3 Feeding Dry Hay

10. SILO CAPACITIES AND FEED DATA
10.1 Silage
     Determining Silo Size
         Silo height
         Silo diameter

11. UTILITIES
11.1 Electrical
     Materials
         Surface mount cable
         Plastic conduit
         Switches, outlets, and junction boxes
         Locating outlets
11.2 Lighting
     Incandescent
     Fluorescent
     High Intensity Discharge (HID)
     Light Levels
     Locating Lights
         Stanchion barn
         Free stall barn
         Pens and stalls
         Milking parlor
         Milk room
         Outside feeding bunks
11.2 Electric Motors
11.3 Service Entrances
11.3 Standby Power
11.3 Water Supply
     Installation
11.4 Lightning
11.4 Electric Fences
     Fence Chargers
     Fence Insulators
     Construction

12. EQUIPMENT PLANS
12.1 Bunks
12.1 Fenceline Bunks
12.3 Covered Bunks
12.4 Lot Bunks
12.5 Feeding Equipment
12.5 Feeding Fences
12.6 Portable Hay Feeder
12.7 Slant Bar Feeder Panels
12.8 Silage Cart
12.9 Mineral Feeders
12.10 Fences
12.10 Windbreak Planning
12.11 Windbreak Fences
12.12 Fencing Corners
12.13 Fencing
12.14 Gates
12.17 Hinges and Latches
12.18 Stock Guards
12.19 Stiles and Passes

12.20 Handling Equipment
12.20 Calf Stall and Pen
12.21 Movable Calf Hutch
12.21 Calf Hay Feeders
12.22 Corrals
12.23 Loading Chutes
12.24 Sunshades
12.25 Calf Shelter
12.26 Super Calf Hutch
12.27 Working Chute

13. CONCRETE
13.1 Concrete Strength and Durability
13.1 Air-Entrained Concrete
13.1 Construction
13.2 Curing
13.2 Slip Resistant Concrete Floors

14. SELECTED REFERENCES

15. INDEX

1. DATA SUMMARY

Table 1-1. Cow stall platform sizes.
Use electric cow trainers. Dimensions from edge of curb to edge of gutter.

Cow weight	Stanchion stalls Width	Length	Tie stalls Width	Length
Under 1,200 lb	4'-0"	5'-6"	4'-0"	5'-9"
1,200-1,600 lb	4'-6"	5'-9"	4'-6"	6'-0"
Over 1,600 lb	Not recommended		5'-0"	6'-6"

Table 1-2. Recommended stall barn dimensions.

Alley width
Flat manger-feed alley	5'8"-6'6"
Step manger-feed alley	6'0"-6'6"
Step manger (24")	
Feed alley (4'0"-4'6")	
Service alley with barn cleaner	6'0"
Cross alley[a]	4'6"

Manger width
Cows under 1,200 lb	20"
Cows 1,200 lb or more	24"-27"

Gutters
Width[b]	16" or 18"
Depth, stall side	11"-16"
Depth, alley side	11"-14"

[a] Taper the end stalls inward 6" at the front for added turning room for a feed cart.
[b] Or as required for barn cleaner.

Table 1-3. Free stall dimensions.
Stall width measured center-to-center of 2" pipe dividers. For wider divider dimensions, increase stall width accordingly. Stall lengths are measured from front of stall to alley side of curb.

Heifers	Width	Length
5-8 mo	2'-6"	5'-0"
9-12 mo	3'-0"	5'-6"
13-15 mo	3'-6"	6'-6"
16-24 mo	3'-6"	7'-0"
Cows (average herd weight)		
1,000 lb	3'-6"	6'-10"
1,200 lb	3'-9"	7'-0"
1,400 lb	4'-0"	7'-0"
1,600 lb	4'-0"	7'-6"

Table 1-4. Typical free stall alley widths.

Feeding and stall access alley	10'-12'
Access alley between 2 stall rows	
Solid floor	8'-10'
Slotted floor	6'-9'
Feeding alley	9'-10'

Table 1-5. Replacement animal space requirements.

1-5a. Calf housing.

Housing type	Pen size
0-2 mo (individual pens)	
Calf hutch (plus 4'x6' outdoor run)	4'x8'
Bedded pen	4'x7'
Tie stall	2'x4'
3-5 mo (groups up to 6 head)	
Super calf hutch	25-30 ft²/hd
Bedded pen	25-30 ft²/hd

1-5b. Heifer housing.

Housing type	5-8	9-12	13-15	16-24
	ft²/animal			
Resting area and	25	28	32	40
paved outside lot	35	40	45	50
Total confinement				
Bedded resting area*	25	28	32	40
Slotted floor	12	13	17	25

*Assume access to 10' wide scraped feed alley.

Table 1-6. Bunk design.

Throat height (max.)	
Calves	18"
Heifers	20"
Mature cows	24"
Bunk width (max. 60")	
Both sides feeding	
Calves	36"
Heifers	48"-60"
Mature cows	48"-60"
One side feeding	18" bottom width
Mechanical feeder	Add 6"-12" up to max. width
Step along bunk	
Height	4"-6"
Width	12"-16"
Bunk apron	
Slope	¾"-1"/ft
Width	10'-12'
Neck rails	
⅜" cable, 2" pipe, 2x6 plank	16"-24" opening

Table 1-7. Feeding space requirements.

	3-4	5-8	Age, months 9-12	13-15	16-24	Mature cow
	in/animal					
Self feeder						
Hay or silage	4	4	5	6	6	6
Mixed ration or grain	12	12	15	18	18	18
Once-a-day feeding						
Hay, silage, or ration	12	18	22	26	26	26-30

Table 1-8. Floor and lot slopes.

Handling facilities	¼"-½"/ft
Lots	
Paved	⅛"/ft min.
Earth	½"-¾"/ft
Mound sideslope	1'/5'
Bunk apron	¾"-1"/ft nearly self-cleaning
	½"/ft min.

Table 1-9. Water requirements.

	gal/hd/day
Calves	
(1-1.5 gal/100 lb)	6-10
Heifers	10-15
Dry cows	20-30
Milking cows	35-45

Table 1-10. Dairy barn ventilating rates.
Size the system based on total building capacity. Table values are additive—e.g. for calves, mild weather requires 15 + 35 = 50 cfm/calf.

	Cold weather*	Mild weather	Hot weather
		cfm/animal	
Calves 0-2 mo	15	+35 = 50	+50 = 100
Heifers			
2-12 mo	20	+40 = 60	+70 = 130
12-24 mo	30	+50 = 80	+100 = 180
Cow 1,400 lb	50	+120 = 170	+300 = 470
Milkroom			600 cfm
Milking parlor		100 cfm/stall	400 cfm/stall

*An alternative cold weather rate is 1/15 the room or building volume; ft³/15. An alternate hot weather rate is the building volume divided by 1.5.

Table 1-11. Dairy manure production.
Provide 2.5 ft³/day of storage per 1,000 lb liveweight for solid manure with bedding. Table values based on manure at 87.3% water and 62 lb/ft³.

Animal size, lb	Total manure production			Nutrient content		
	lb/day	ft³/day	gal/day	N	P	K
					lb/day	
150	12	0.19	1.5	0.06	0.010	0.04
250	20	0.33	2.4	0.10	0.020	0.07
500	41	0.66	5.0	0.20	0.036	0.14
1,000	82	1.32	9.9	0.41	0.073	0.27
1,400	115	1.85	13.9	0.57	0.102	0.38

Table 1-12. Typical herd makeup.
Replacement numbers assume uniform calving year-round, 12-month calving interval, no death loss or culling, 50% male and 50% female calves, and all males sold at birth.

	Number				Avg. weight lb
Cows milking	33	62	83	208	1,400
Dry cows	7	13	17	42	1,550
Herd size = total mature cows	40	75	100	250	1,450
Heifers					
16-24 mo	15	28	38	95	1,050
13-15 mo	5	9	12	30	800
9-12 mo	7	13	17	43	600
5-8 mo	7	13	17	42	400
3-4 mo	3	6	8	20	250
Calves 0-2 mo	3	6	8	20	150
Total replacements	40	75	100	250	

Table 1-13. Conversions.
Multiply to the right: acres x 43,560 = ft².
Divide to the left: ft² ÷ 43,560 = acres.

Unit	Times	Equals
Acres	43,560	ft²
	4,840	yd²
	160	square rods
	1/640	square mile
Acre-ft	325,851	gallons
	43,560	ft³
Acre-in	3,630	ft³
Acre-in/hr	453	gpm
	1	cfs (approximate)
Bushels	1.25	ft³
	2.5	ft³ ear corn
ft³	7.48	gallons
	1728	in³
	62.4	lb water
	0.4	bu ear corn
	0.8	bu grain
cfs	448.8	gpm
	646,317	gal/day
Cubic yard	27	ft³
concrete	81	ft² of 4" floor
concrete	54	ft² of 6" floor
Gallons	231	in³
	0.134	ft³
	8.35	lb water
Miles	5,280	ft
	1,760	yd
	320	rods
Pressure, psi	2.31	ft of water head
Rods	16.5	ft
	5.5	yd

2. TOTAL DAIRY FACILITY

When planning new construction or major modification of a dairy system, consider:
- Calf, heifer, dry cow, and milking cow housing.
- Feed types, handling equipment, and storage.
- Manure handling method.
- Milking system and equipment.
- Labor requirements.
- Building environment.
- Sanitary and pollution control regulations.
- Future expansion.

Many dairy farmers produce their own feeds and raise their own herd replacements. The needs of each group require different housing, feeding, storage, and handling systems.

Herd Makeup

Herd size can mean either the number of cows actually milking or the number of both dry and milking cows. In this book, herd size is the total of dry and milking cows. Add calves and heifers to the herd makeup if you raise replacements. Table 2-1 gives typical herd makeups assuming uniform calving year-round.

Table 2-1. Typical herd makeup.
Replacement numbers assume uniform calving year-round, 12-month calving interval, no death loss or culling, 50% male and 50% female calves, and all males sold at birth.

	Number				Avg. Weight lb
Cows milking	33	62	83	208	1,400
Dry cows	7	13	17	42	1,550
Herd size = total mature cows	40	75	100	250	1,450
Heifers					
16-24 mo	15	28	38	95	1,050
13-15 mo	5	9	12	30	800
9-12 mo	7	13	17	43	600
5-8 mo	7	13	17	42	400
3-4 mo	3	6	8	20	250
Calves 0-2 mo	3	6	8	20	150
Total replacements	40	75	100	250	

Feeds and Cropland

Determine the best ration for each group of animals based on available feeds, feed quality, animal size, and milk production levels. Estimate cropland and storage needs based on your ration and total number of animals. Without a specific ration, use Table 2-2 to determine approximate storage needs.

Cropland needed is affected by milk production, ration, forage choice, crop yields, etc. See Table 2-3 for estimated cropland needs. If all feeds except supplements are raised on the farm, a good estimate is 3 to 4 acres per cow and replacement.

Table 2-2. Annual feed requirements.
Table values are for each cow and replacement. Determine total needs based on herd size. If using high moisture grain, multiply shelled corn figures by 1.267 to get bushels based on 30% moisture grain and a density of 60 lb/bu (1.25 ft³/bu).

	% dry matter	Lb milk/cow-year			
		12,000	14,000	16,000	18,000
		Quantities/cow and replacement			
Medium level of silage:					
Hay silage, T	40-50	11.0	11.6	12.3	12.9
or hay, T	80-85	4.3	4.5	4.7	4.9
Corn silage, T	30-35	12.0	12.0	12.0	12.0
Shelled corn, bu	84.5	72	84	102	118
High level of silage:					
Hay silage, T	40-50	6.7	6.9	7.1	7.5
or hay, T	80-85	2.6	2.7	2.8	2.9
Corn silage, T	30-35	16.5	16.5	16.5	16.5
Shelled corn, bu	84.5	52	64	86	106

Source: *Chore Reduction for Free Stall Dairy Systems, Hoard's Dairyman*, Fort Atkinson, WI.

Table 2-3. Estimated cropland.
Acreage to produce annual feed required per cow and replacement. If using high moisture grain, adjust shelled corn figures. Values based on Table 2-2 and yields shown.

	% dry matter	Lb milk/cow-year			
		12,000	14,000	16,000	18,000
		acres/cow and replacement			
Medium level of silage:					
Hay silage 6 T/A	40-50	1.8	1.9	2.1	2.2
or hay 3 T/A	80-85	1.4	1.5	1.6	1.6
Corn silage 15 T/A	30-35	0.8	0.8	0.8	0.8
Shelled corn 80 bu/A	84.5	0.9	1.1	1.3	1.5
High level of silage:					
Hay silage	40-50	1.1	1.2	1.2	1.3
or hay	80-85	0.9	0.9	0.9	1.0
Corn silage	30-35	1.1	1.1	1.1	1.1
Shelled corn	84.5	0.7	0.8	1.1	1.3

Housing

Provide housing for different animal groups based on Table 2-1. More than one group can be housed in the same building, but allow for managing each group separately. Also allow for different requirements for sanitation, environment, etc. In larger dairies, separate facilities may be provided for each group.

Farmstead Planning

Many factors determine the best plan, and while some are common sense, overlooking one can cause a poorly planned farmstead. Collect ideas from publications, farm visits, county agents, experienced producers, scientists, and engineers. Plan on paper, where mistakes can be easily corrected. It is less costly to correct a mistake during the planning stage than after construction begins. Stake out the best arrangements on the site to see how they fit.

Consider the entire farmstead when planning a new or modified housing system. Solving one problem may create another. With proper planning and attention to details, a well organized, functional farmstead can result.

Site Selection

Many factors determine the best site for dairy facilities. Provide for these items during construction.

Space for buildings, clearance between buildings (at least 35′ for most buildings and 50′ for naturally ventilated buildings), lots, and expansion. Assume the operation will double in size and plan accordingly. Provide lanes for vehicle access and room for parking. Allow for a feed center and adequate separation from family housing, Fig 2-1. A typical 100-cow dairy farmstead requires 2 to 6 acres for barns, lots, home, machinery, and feed storage.

Drainage away from barns and lots. Ditch and fill low areas. Divert runoff away from buildings and traffic areas. Provide a 2%-5% slope on outside lots. Use mounds to provide dry resting areas. Earth moving is inexpensive compared to facility costs.

Wind and snow control. Windbreaks help deflect winter winds and control snow. Take advantage of trees, buildings, hills, and haystacks for winter wind protection. Allow for summer air movement and drainage when locating windbreaks. Consider prevailing wind directions for reducing odors, snow drifting, insects, and noise.

Water. A year-round supply of potable water is essential for watering animals and sanitation. Water is also needed for fire protection and waste dilution.

Milking cows need 35 to 40 gal/head-day (4½ to 5 lb water/lb milk produced). Peak water consumption is shortly after feeding. Provide a system that meets peak and total daily requirements. Where ground water supplies are not adequate, use surface sources such as farm ponds or community water systems. Approval of your water system may be required. For more information see MWPS-14, *Private Water Systems Handbook*.

Access. Provide all-weather roads for milk trucks, repair persons, technicians, veterinarians, feed handling equipment, etc. Provide adequate parking for visitors. Minimum road width is 12′. Minimum turning radius for large milk trucks is about 55′. A hay wagon can turn 180° in about 50′.

Manure storage, handling, and disposal. Select a site with sufficient land for spreading manure. Minimum acreage required by many state pollution agencies is based on satisfying the nitrogen requirement of the growing crop.

Avoid steep slopes where manure runoff can cause water pollution, and avoid land adjacent to neighboring residences.

Feed storage and handling is a consideration, but not an overriding concern, in choosing a site. Transport wagons can link a grain-feed center to dairy facilities.

Electric power is needed for heating, lighting, pumps, and motors. A 200 amp, 220 volt barn entrance is common. Thorough grounding reduces stray current problems. Provide standby emergency power in the event of a power outage. Some producers install three phase power; consult your power supplier.

Security. Consider theft, vandalism, and fire safety. Limit farm visitor access to control disease and to reduce interference with farm work. If located on the same farmstead as the manager's residence, run the access lane near the home. If a second access is used for feed, manure, and animal transport vehicles, provide an alarm system to guard against unauthorized traffic.

Facilities remote from the manager's residence pose the most problems. Provide only one access road—unauthorized persons are less apt to visit if there is no escape route should the manager return. If possible, make access roads at remote sites visible from a public road or neighboring residence.

Remodeling

When planning to expand, you may have to decide whether to remodel or abandon an existing building. Carefully consider future as well as present needs. Evaluate these general factors:
- Compatibility with final setup.
- Structural integrity.
- Location of existing building.
- Cost of remodeling vs. new building.

Remodeling is not always the cheaper route, especially when future needs are considered. If remodeling cost is more than ½ to ⅔ new building cost, a new building is usually best. Sometimes, it is possible to use some materials from an existing building in a new one.

Plans, Specifications, and Contracts

Detailed documents help provide needed communication and understanding between owner and builder. **Plans** show all necessary dimensions and details for construction. **Specifications** support the plans; they describe the materials to be used, including size and quality, and often outline procedures for construction and quality of workmanship. The **contract** is an agreement between the builder and the owner; it includes price of construction, schedule of payments, guarantees, responsibilities, and starting and completion dates.

You have several options for preparing this material.

Be your own contractor. Draw a final plan, making sure dimensions are correct and construction details and materials are determined. Have plans checked by the appropriate regulatory agency when required. Determine total costs before beginning construction.

Hire a consulting engineer.

Use a design and construction firm. Some firms make working drawings and specifications and have standard contract forms.

Consulting Engineers

Services commonly offered by consulting engineers include:
- Direct personal service (technical advice, etc.).
- Preliminary investigations, feasibility studies,

and economic comparison of alternatives.
- Planning studies.
- Design.
- Cost estimates.
- Engineering appraisals.
- Bid letting.
- Construction supervision and inspection.

Consulting engineers usually do a project in three phases: preliminary planning, engineering design, and construction monitoring. They may be retained to help with one or more of these phases. To select a consulting engineer, consider:

- Registration: to protect the public welfare, states certify and license engineers of proven competence. Practicing consulting engineers must be registered professional engineers in their state of residence, and qualified to obtain registration in other states where their services are required.
- Technical qualifications.
- Reputation with previous clients.
- Experience on similar projects.
- Availability for the project.

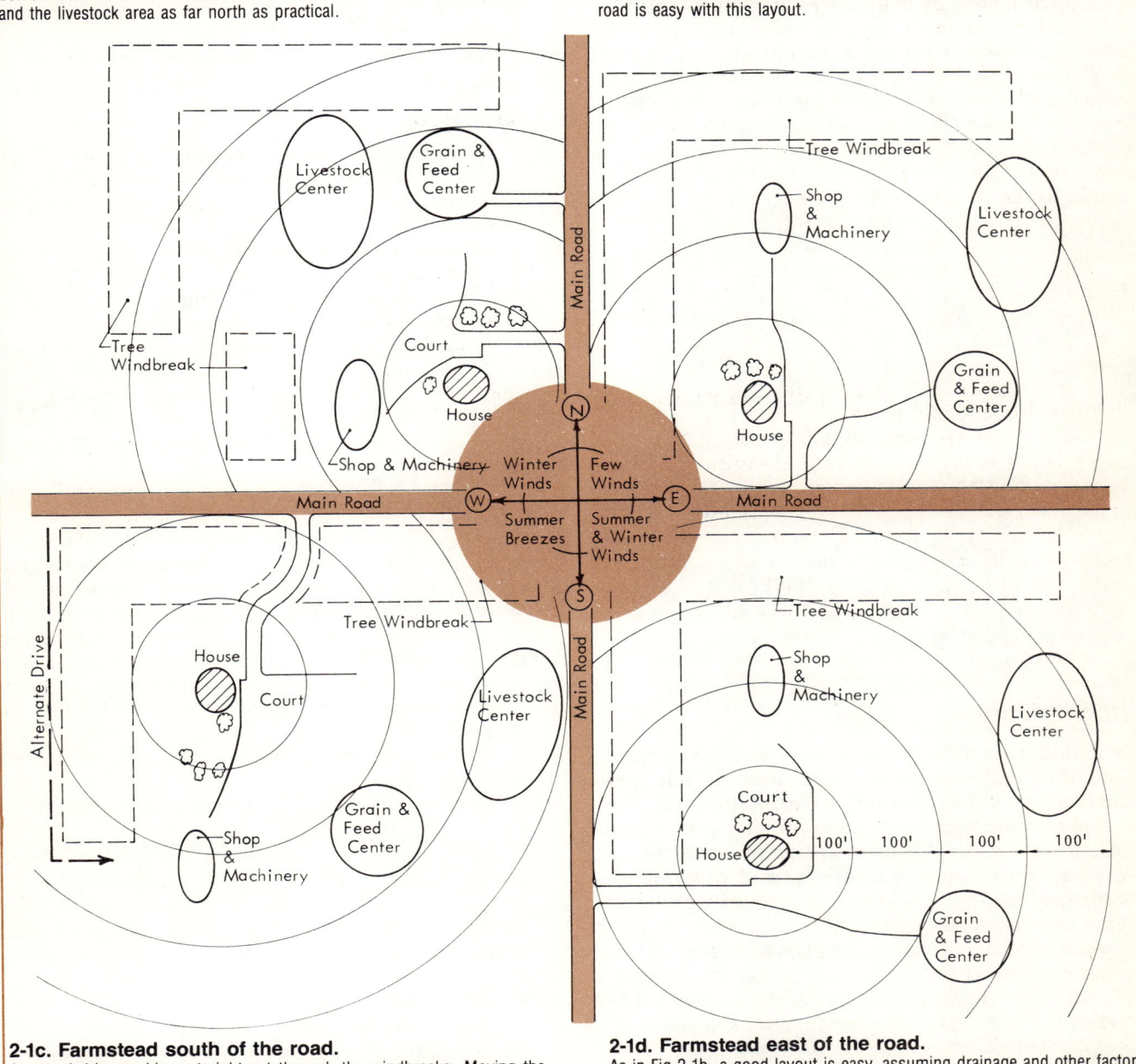

2-1a. Farmstead west of the road.
Some winter winds come from the northwest. Locate the house as far west and the livestock area as far north as practical.

2-1b. Farmstead north of the road.
A good relationship between house, windbreak, livestock center, and main road is easy with this layout.

2-1c. Farmstead south of the road.
A curved drive avoids a straight cut through the windbreaks. Moving the house farther south and the livestock area northeast is desirable. If the house and machine center can be reversed, use the alternate drive.

2-1d. Farmstead east of the road.
As in Fig 2-1b, a good layout is easy, assuming drainage and other factors permit this arrangement.

Fig 2-1. Farmstead and main road relationships.
The direction of the farmstead from the main road affects farmstead layout. See MWPS-2, *Farmstead Planning Handbook,* for additional information on site planning.

3. REPLACEMENT ANIMAL HOUSING

Consider housing replacement animals away from the milking herd. Separating replacement animals into groups according to age or size allows each group to be treated according to its needs. Suit space, equipment, environment, rations, and care to the age group being handled. When planning replacement animal housing, provide:
- Adequate space for animal groups based on size or age.
- Adequate feed and water space.
- Dry, bedded, covered, draft-free resting areas.
- Fresh air.
- Clean lots and proper sanitation.

Well designed housing promotes healthier replacements and more efficient labor use. Neglecting these factors can stress the animal and increase the chance of disease or injury.

Management

Good management is important to the success of any housing system. Management is your understanding of what to do and when to do it. Sanitation, ventilation, feeding, treatment, and close observation are all important management practices.

For healthy, potentially high producing replacement animals, consider the housing environment. Provide adequate space for water, feed, resting, and exercise.

Space Requirements

The number of replacement animals is proportional to the number of milking cows in the herd, Table 2-1. As herd size expands, replacement numbers increase. Increase space for these animals to avoid crowding. Injuries can increase under crowded conditions. See Table 3-1 for minimum space requirements.

Environment Alternatives

A **cold barn** has indoor winter temperatures about the same as outdoors. It is usually uninsulated and has unregulated natural ventilation.

Cold barns are the least expensive to build, but freezing and condensation on inside walls during severe winter weather can be problems. Heated waterers are needed. Adjust rations to maintain sufficient body heat production.

A **modified environment barn** has indoor winter temperatures higher than outdoors and usually above 32 F. A modified barn is insulated and designed for controlled natural or fan ventilation.

A modified barn has fewer problems with frozen manure. Insulation increases the cost of modified housing but helps keep the barn cooler in summer and warmer in winter. Ventilation, natural or mechanical, must be properly managed.

A **warm barn** has uniform inside winter temperatures of 40 F or above. It has insulation and mechanical ventilation and can use supplemental heat for environmental control. Warm barns have more comfortable working conditions but are more expensive to build and operate.

Table 3-1. Replacement animal space requirements.

3-1a. Calf housing.

Housing type	Pen size
0-2 mo (individual pens)	
Calf hutch (plus 4'x6' outdoor run)	4'x8'
Bedded pen	4'x7'
Tie stall	2'x4'
3-5 mo (groups up to 6 head)	
Super calf hutch	25-30 ft²/hd
Bedded pen	25-30 ft²/hd

3-1b. Heifer housing.

Housing type	5-8	9-12	13-15	16-24
	\-\-\-\-\-\-\-\- ft²/animal \-\-\-\-\-\-\-\-			
Free stall	2'6"x5'	3'x5'6"	3'6"x6'6"	3'6"x7'0"
Manure alley width	8'-10'	8'-10'	8'-10'	8'-10'
Resting area and	25	28	32	40
paved outside lot	35	40	45	50
Total confinement				
Bedded resting area*	25	28	32	40
Slotted floor	12	13	17	25

*Assume access to 10' wide scraped feed alley.

Housing Types

Bedded pack
- Bedding is required.
- Manure is stored in barn (clean out every 3 to 6 mo).
- Pack provides a warm resting surface.

Free stalls (solid floor)
- Small amount of bedding required.
- Frequent manure removal (unless slotted floor is provided).
- Frozen manure can be a problem in "cold" barns.

Totally slotted floor
- No bedding required.
- Under-building manure storage (clean out every 6 to 8 mo).
- Adequate ventilation needed because of manure gases.
- Drafts and/or cold temperatures can be a problem with young heifers (2 to 6 mo).

Counter-slope floor
- No bedding required.
- Frequent manure removal, except with slotted floor in litter alley.
- Maintain high stocking rates to keep animals cleaner.
- Control runoff from uncovered litter alleys to prevent stream and groundwater pollution.
- Not recommended for bred heifers.

Calf Housing

Calf housing and management depend on age. The sucking instinct persists in unweaned calves (0 to 2 mo) for about 30 min after feeding. Individually pen calves, because mouth-to-mouth contact can transmit disease and calf sucking can cause udder damage. House unweaned calves separately from the milking herd and maternity area.

With cold housing systems, 4'x7' bedded pens with 4' high partitions work well, Fig 3-1. Solid partitions minimize drafts and calf-to-calf contact. Provide a combined feed and manure alley.

Unweaned calves grow well in outdoor calf hutches, Fig 3-2. Hutches are individual, open-front shelters usually with a 4'x6' outside run. They provide semi-isolation for individual animals. Separate hutches to prevent facial contact through the fence of the outside run. Move hutches after each use to improve sanitation. Provide winter wind protection.

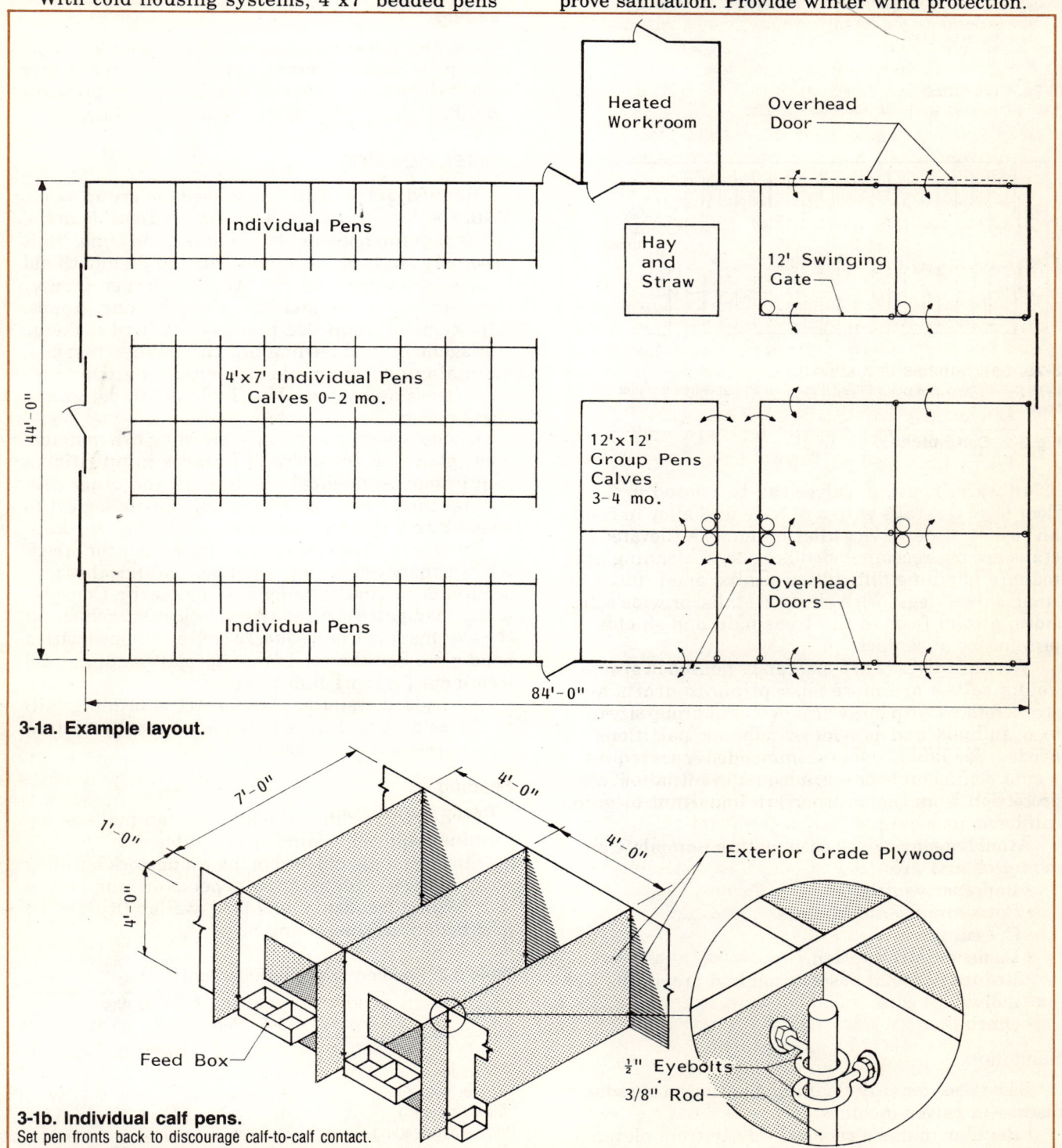

3-1a. Example layout.

3-1b. Individual calf pens.
Set pen fronts back to discourage calf-to-calf contact.

Fig 3-1. Naturally ventilated calf barn.
Use solid pen sides to reduce drafts. Provide wall openings near the floor for summer air movement. Limit group size to 5 to 6 calves.

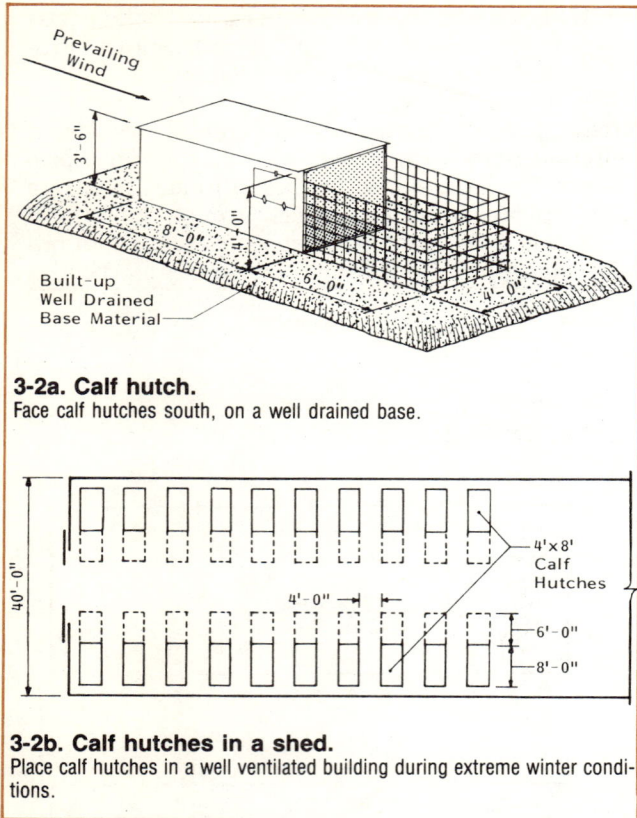

3-2a. Calf hutch.
Face calf hutches south, on a well drained base.

3-2b. Calf hutches in a shed.
Place calf hutches in a well ventilated building during extreme winter conditions.

Fig 3-2. Calf hutches.

In warm housing, calves can be housed in 2'x4' floor level tie stalls with a 4' wide feed alley in front and a 3'-4' wide service alley in the rear. Elevated tie stalls are not recommended because of cleaning and manure handling difficulties, drafts, and bruises to small calves' legs. With elevated stalls, provide solid sides, a solid floor in the front half, and shields to prevent facial contact.

Weaned calves (2 to 5 mo) can be housed in groups. Young calves are more susceptible to drafts and stress related with large groups. Limit group size to 5 to 6 animals and provide windbreak partitions if needed. See Table 3-1 for recommended space requirements. Sufficient bedding, adequate ventilation, and protection from the weather are important to good calf housing.

Avoid housing calves with mature animals. Common problems are:
- Improper ventilation.
- Cows are disease carriers.
- Crowding.
- Inadequate sanitation.
- Reduced animal observation and provisions for individual care.
- Drafts.

Sanitation

Build pens for easy cleaning. Sanitation to reduce disease in calves includes:
- Regular manure removal—daily from elevated stalls.
- Washing stalls, pens, and hutches after animal removal.
- Resting or drying out pens and hutches for at least a week between use, which requires more facilities.
- Moving hutches to a new site to break disease cycles.

Put newborn calves on fresh bedding and add bedding as needed to keep them dry.

Feeding

Locate a workroom near calf housing for feed storage, milk replacer preparation, and a hot water heater. Hot water is needed for milk replacer preparation. Provide for storing calf starter and hay.

Heifer Housing

Weaned calves can be housed in group pens. Young heifers (2 to 5 mo) are more susceptible to the stress of group housing. For heifers up to 5 mo, limit group size to 5 to 6 animals. Six- to 24-month-old heifers can withstand the stress of larger groups. Consider heifer age and size when forming groups. Replacement group size is related to herd makeup, management, and feeding practices. When possible, maintain uniform animal size within a group to reduce stress and injury. Use Table 3-1 to determine space needs.

Divide heifers into groups according to a management plan that considers differences in nutritional requirements, medical treatments and other procedures, and breeding. For example, it is logical to transfer heifers to the next group when they are bred.

Provide facilities for restraining heifers for breeding, pregnancy-checking, vaccinations and other procedures either in the heifer barn or nearby. Consider using head-gates in cross alleys, lock-in stanchions at the feed manager, or a squeeze chute equipped with a head-gate. See the section on special handling and treatment for more information.

For bedded systems, plan for ½ T of bedding/calf-yr. If bedding is limited, consider totally slotted floors, free stalls, or sloped floors.

Feeding

Feed type, feeding schedule, and animal size determine required feeding space, Table 3-2.

Store bedding and feed in the youngstock building when possible. Storage space depends on animal density, feeding frequency, and feed availability. Do not feed heifers in the resting area.

Table 3-2. Feeding space requirements.

	\multicolumn{5}{c}{Age, months}
	3-4	5-8	9-12	13-15	16-24
	\multicolumn{5}{c}{in/animal}
Self feeder					
Hay or silage	4	4	5	6	6
Mixed ration	12	12	15	18	18
Once-a-day feeding					
Hay, silage, or ration	12	18	22	26	26

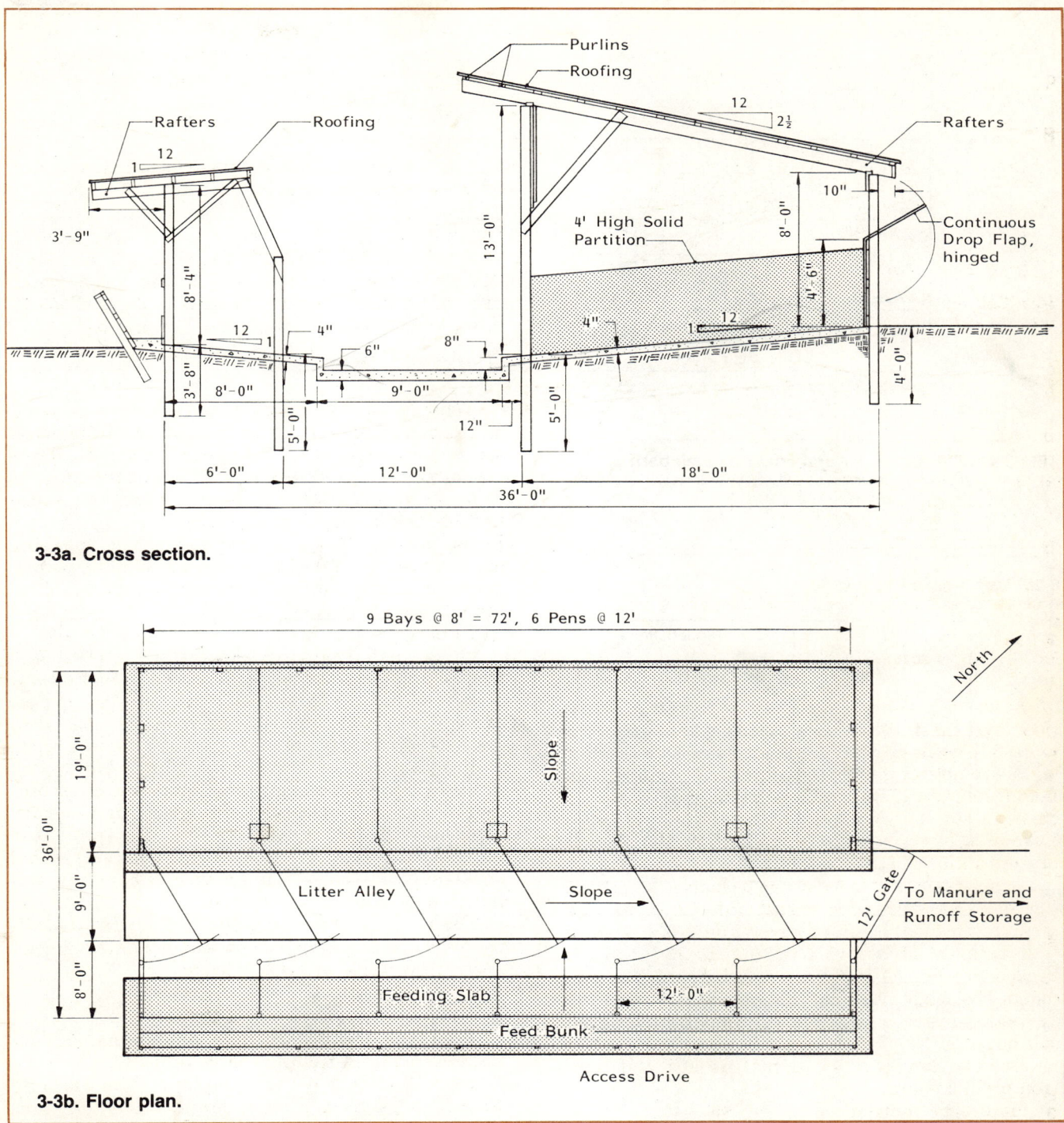

3-3a. Cross section.

3-3b. Floor plan.

Fig 3-3. Counter-slope youngstock barn.
Sample barn layout for 72 to 90 animals with outdoor feeding and scraped manure alley.

Bred Heifers and Dry Cows

First calf heifers and dry cows are often housed together, especially when heifers are within three months of calving. Consider housing them separately from the milking herd. Tie stalls are expensive; free stalls are adequate.

Observe bred heifers and dry cows frequently for calving signs and move them to maternity facilities before calving. Locate the maternity area near the dry cow and bred heifer group for easy access.

Figures 3-4 and 3-5 show two types of housing for heifers and dry cows. The building layout and dimensions are similar. The difference between housing types is the resting area. It is either bedded pack or free stalls. With some planning, free stalls could be added later to the bedded pack building.

3.5

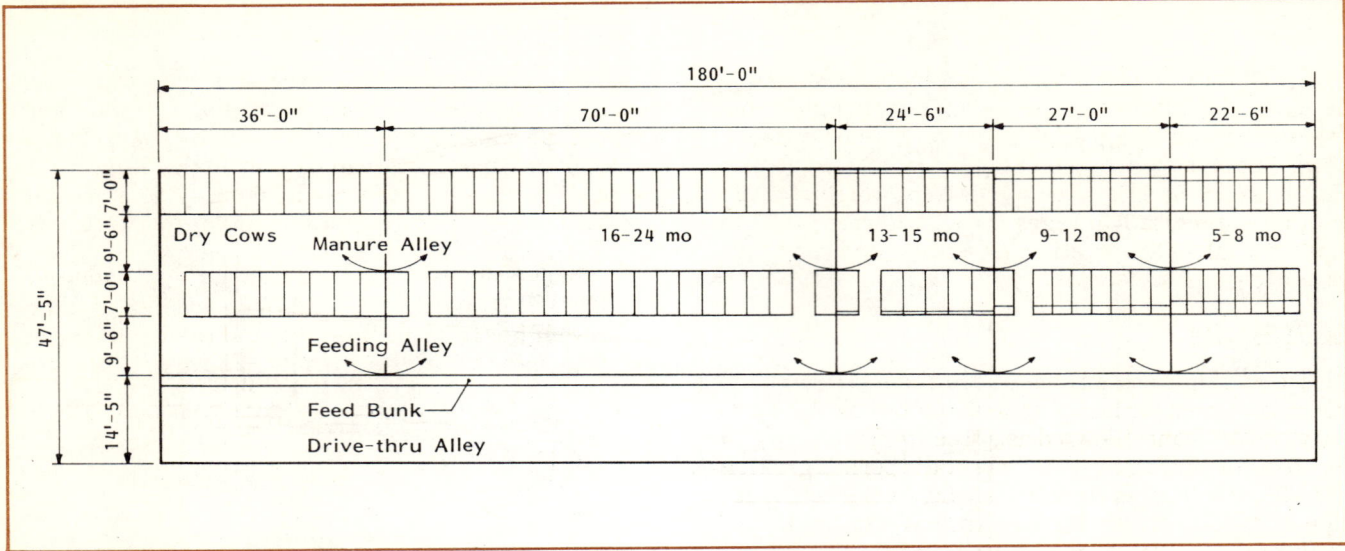

Fig 3-4. 102-head heifer and cow free stall barn.
A free stall barn for 4 groups of heifers and a group of dry cows. An alternative free stall plan with different group sizes is available, mwps-72343.

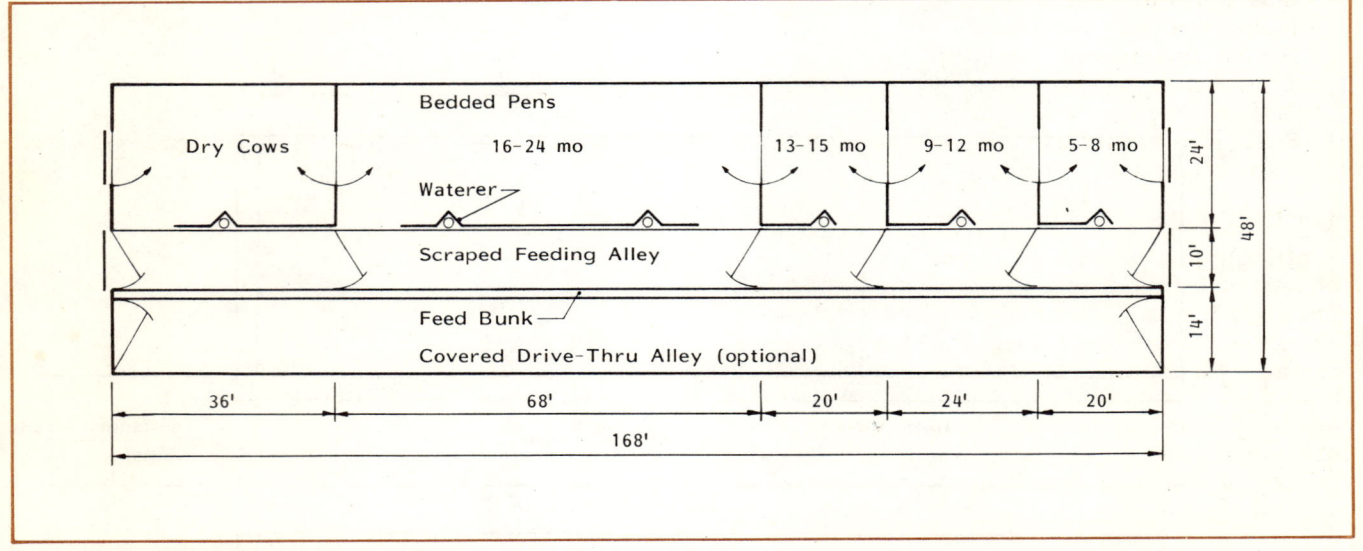

Fig 3-5. Bedded pack heifer and cow barn.
A bedded pack barn for 4 groups of heifers and a group of dry cows. Gates across the feeding alley are used to hold animals in bedded pens during alley scraping.

4. MILKING HERD FACILITIES

Two types of housing are common for milking herds in the Midwest—tie stall (or stanchion) and free stall barns. Some loose housing systems are still used, but most have been converted to free stalls. Before deciding on a housing system, weigh the advantages and disadvantages of each alternative, consider initial and operating expenses, and evaluate your personal preference.

Stall Barns

Tie stall (stanchion) barns are the most common housing system for herds up to about 60 cows. A stall barn can be as mechanized as a free stall barn, except for stooping during milking. Tie stalls permit greater individual attention than free stalls. Initial construction costs for comparable free stall and tie stall systems are about the same up to 60 head.

Before you build or remodel a barn, determine:
- Number of cows. Base milking herd size on farm size, available labor, and production goals.
- Stall arrangement, size, and type.
- Stall floor covering or bedding type.
- Number of pens needed.
- Insulation and ventilation type.
- Type of milking system.
- Manure management system.
- Feed type, handling equipment, and storage.
- Expansion possibilities.

Stanchion Stalls

Stanchion stalls have a metal yoke fastened at the top and bottom that is free to swing from side to side, Fig 4-1. Usually, cows are released individually. With some stalls a lever opens and closes a group of stanchions to reduce labor. On some, the yoke adjusts forward and backward to help position cows so that manure falls in the gutter. These adjustments have little value. Use proper platform size, well adjusted cow trainers, and good management to keep platforms clean.

Water bowls are usually between every other cow over the manger. The water line can be in the floor, part of the stanchion frame, or overhead. Lever stalls without cow partitions can be used for milking stalls with small free stall or loose housing systems. Stalls are also effective in hospital or maternity areas.

Tie Stalls

With tie stalls, each cow has a neck chain or strap to prevent her from backing out, and a front restraint to keep her out of the manger, Fig 4-2. Cows are

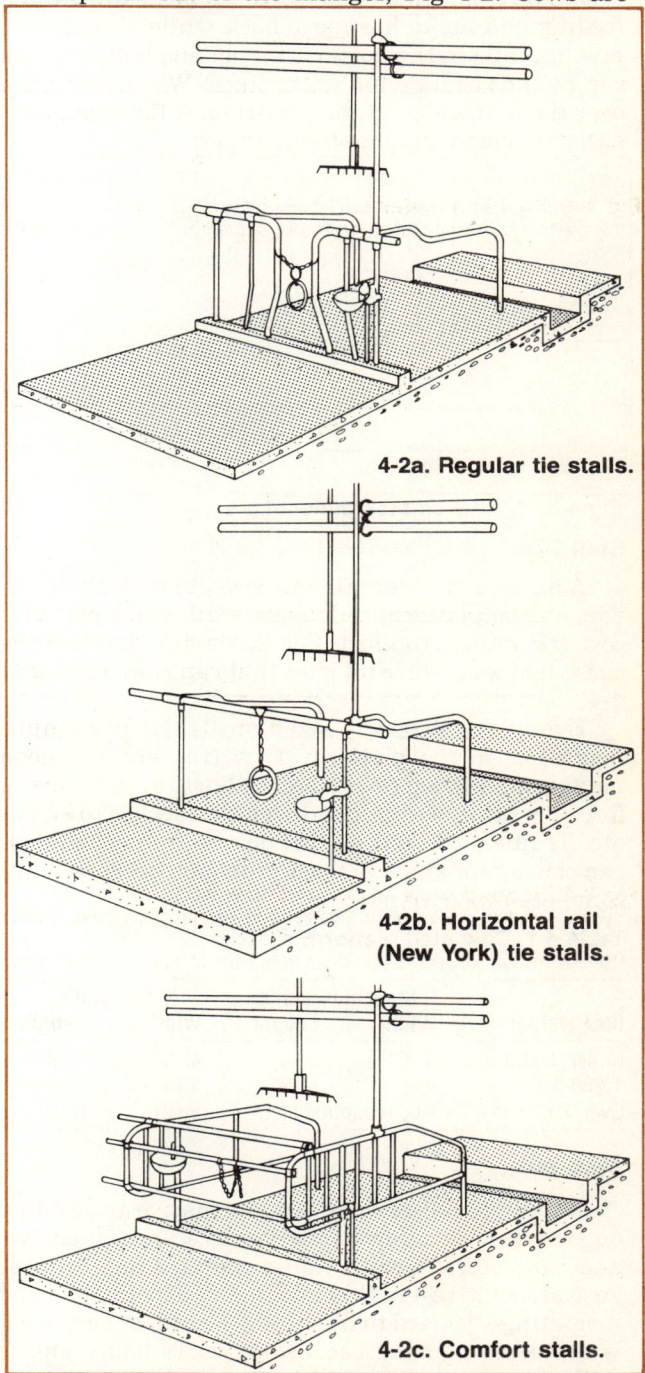

4-2a. Regular tie stalls.

4-2b. Horizontal rail (New York) tie stalls.

4-2c. Comfort stalls.

Fig 4-2. Tie stalls.

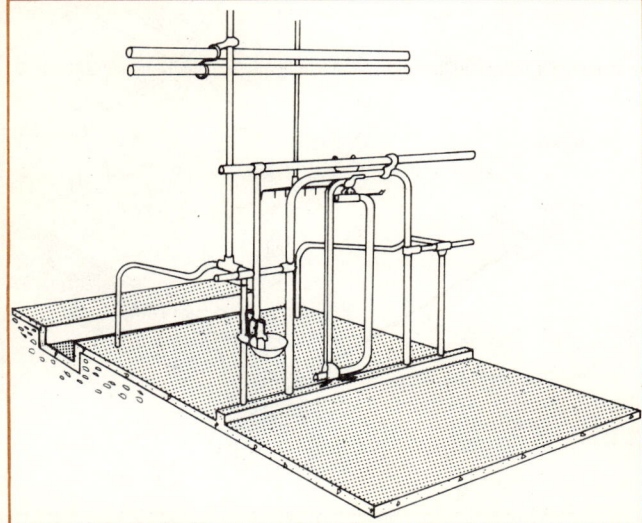

Fig 4-1. Stanchion stalls.
The yoke swings freely from side to side.

fastened and released individually, requiring more labor than with lever stanchions. Generally, tie stalls are preferred over stanchion stalls because of greater cow comfort. Proper stall platform size is important to cow comfort and cleanliness.

Regular tie stalls let the cow lie with her head over the manger or platform.

Horizontal rail stalls have one pipe set in the front curb on each side of the stall. A horizontal pipe, set 7" out over the manger, restrains the cow and can be the water line. Cows have almost completely free head movement.

Comfort stalls are the most expensive tie stall type. The front rails keep the cow's head down when feeding and make her move back while standing so manure falls in the gutter. The top and bottom pipes can be the vacuum and water lines. Water bowls are over the manger or in the partition. A flat manger is easier to clean than a step manger.

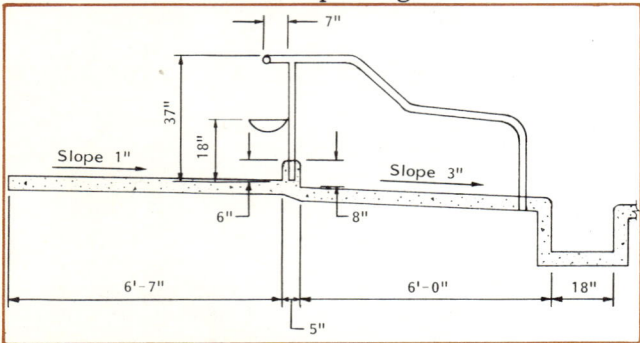
Fig 4-3. Typical tie stall dimensions.

Stall Size

Adequate stall length lets cows lie with their udders on the platform. Adequate stall width provides cow and milker comfort. See Table 4-1. Small stalls cause teat and udder injuries that can ruin a cow, Fig 4-4.

Even with properly-sized stalls, some manure may drop on the platform. Cow trainers and good management keep cows clean. Adjust cow trainers to force cows to move back and arch their backs to evacuate. Trainers located too far on the front shoulders can cause reproductive tract problems because of incomplete evacuation.

Table 4-1. Cow stall platform sizes.
Use electric cow trainers. Dimensions from edge of curb to edge of gutter.

Cow weight	Stanchion stalls Width	Length	Tie stalls Width	Length
Under 1,200 lb	4'-0"	5'-6"	4'-0"	5'-9"
1,200-1,600 lb	4'-6"	5'-9"	4'-6"	6'-0"
Over 1,600 lb	Not recommended		5'-0"	6'-6"

Stall Mats

Many producers want an alternative to bedding on concrete. Bedding is often scarce and expensive, and can interfere with liquid manure handling. A good alternative—highly absorbent, expendable, and insulating—is hard to find. Indoor-outdoor carpeting is not recommended because it wears badly and is difficult to hold in place.

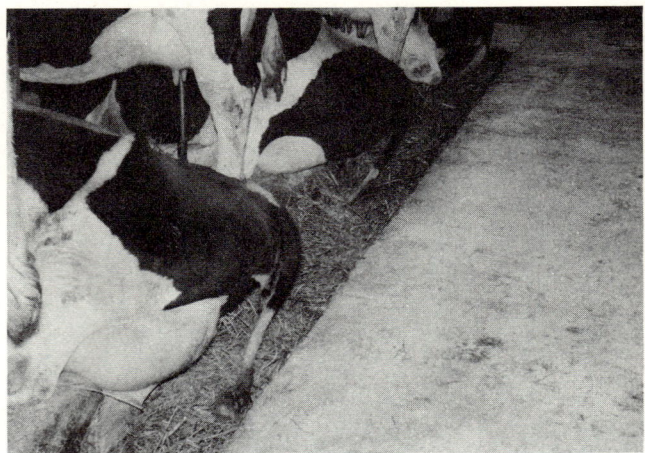
Fig 4-4. Too-short stall platforms.
Udders partly over the gutter are vulnerable to injury. Build the stall platforms long enough.

Rubber mats are common to improve cow comfort. Manure, urine, and bedding accumulate under mats held in place with lag screws; regular cleaning is difficult and time consuming. To reduce these problems, lay rubber mats within one hour of pouring the concrete. Cut each mat smaller than the stall platform by about 6" at the front and back and about 3" along the sides. Cast the alley, manger, and curb with stalls in place. Cast the cow platform concrete to a level even with the **bottom** of the rubber mat. Wait one hour and place the mats. Fill around the mats with concrete finished flush with the mat surface, Fig 4-5. Eventually these mats loosen and are more difficult to clean under than mats placed on hardened concrete.

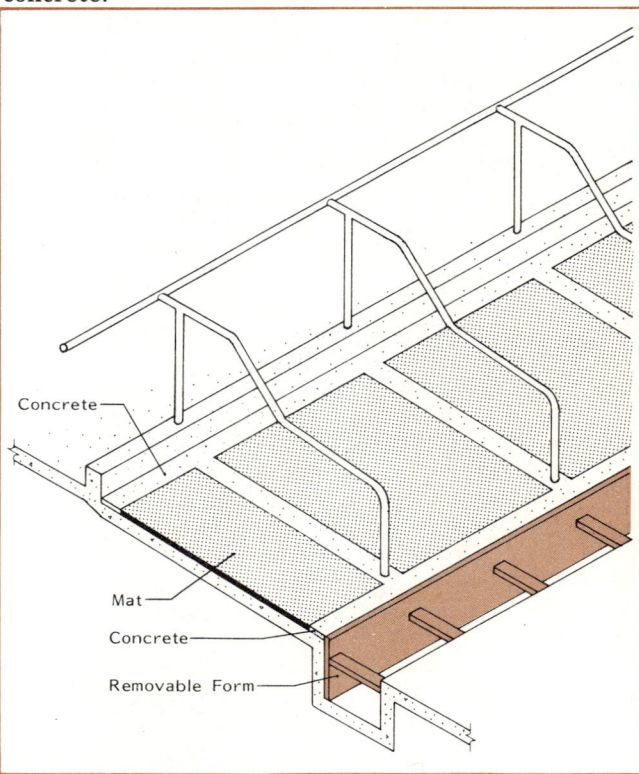
Fig 4-5. Stall mats.

Proper management, not mats, keeps cows clean. Use a little finely-chopped bedding and install cow trainers to make cows deposit manure and urine in a gutter. Because mats insulate and destroy the electrical ground for trainers, use tie chains in tie stalls rather than leather straps. In stanchion stalls, there is usually enough ground through the yoke.

Rubber mats reduce wear on cow hooves, so periodic trimming is required.

Stall Arrangement

An operator spends more time behind the cows than in front. So arrange stalls in two rows facing out for labor efficiency. Only one litter alley is needed and a gutter cleaner can be economical. Walking during milking is reduced and a pipeline milker can be installed more economically.

Although feeding is easier from a common feed alley when cows face in, milking is less convenient with two litter alleys, a pipeline milker is more difficult to install, and litter alley walls become manure spattered.

With a pipeline milker, slope the barn floor about 1"/10' toward the milkhouse, so the milk line height above the floor is nearly constant. Slope gutters 1"/20' or make them flat. Vary gutter depth to prevent manure overflow. Locate the milkhouse at the end of the barn. A milkhouse at the middle of the barn makes it difficult to properly install a pipeline milking system.

Barn Dimensions

Modern stall barns are usually 36' wide outside. A 34' wide barn is sometimes used with medium-size cows. Provide a 6' wide litter alley. Use a flat manger feed alley for easier cleaning and equipment movement, Fig 4-3.

Manure gutters are usually 16" or 18" wide, Fig 4-6. Less urine splashes on the service alley with 18" wide gutters. A depth of 16" on the platform side and 14" on the alley side is common. Some producers prefer the stall and service alley at the same elevation with a 12" gutter. Steel grates over gutters help keep bedding out of liquid manure systems and keep cows' tails cleaner, Fig 4-7.

Slope stall platforms 3" to the gutter.

Table 4-2. Recommended stall barn dimensions.

Alley width
Flat manger-feed alley	5'8"–6'6"
Step manger-feed alley	6'0"–6'6"
Step manger (24")	
Feed alley (4'0"–4'6")	
Service alley with barn cleaner	6'0"
Cross alley[a]	4'6"

Manger width
Cows under 1,200 lb	20"
Cows 1,200 lb or more	24"–27"

Gutters
Width[b]	16" or 18"
Depth, stall side	11"–16"
Depth, alley side	11"–14"

[a]Taper the end stalls inward 6" at the front for added turning room for a feed cart.
[b]Or as required for barn cleaner.

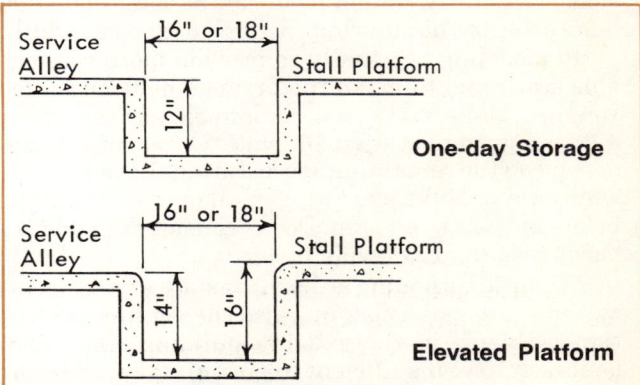

Fig 4-6. Typical barn cleaner gutter.
Common gutter depths are 14" next to the service alley and 16" next to the stall platform.

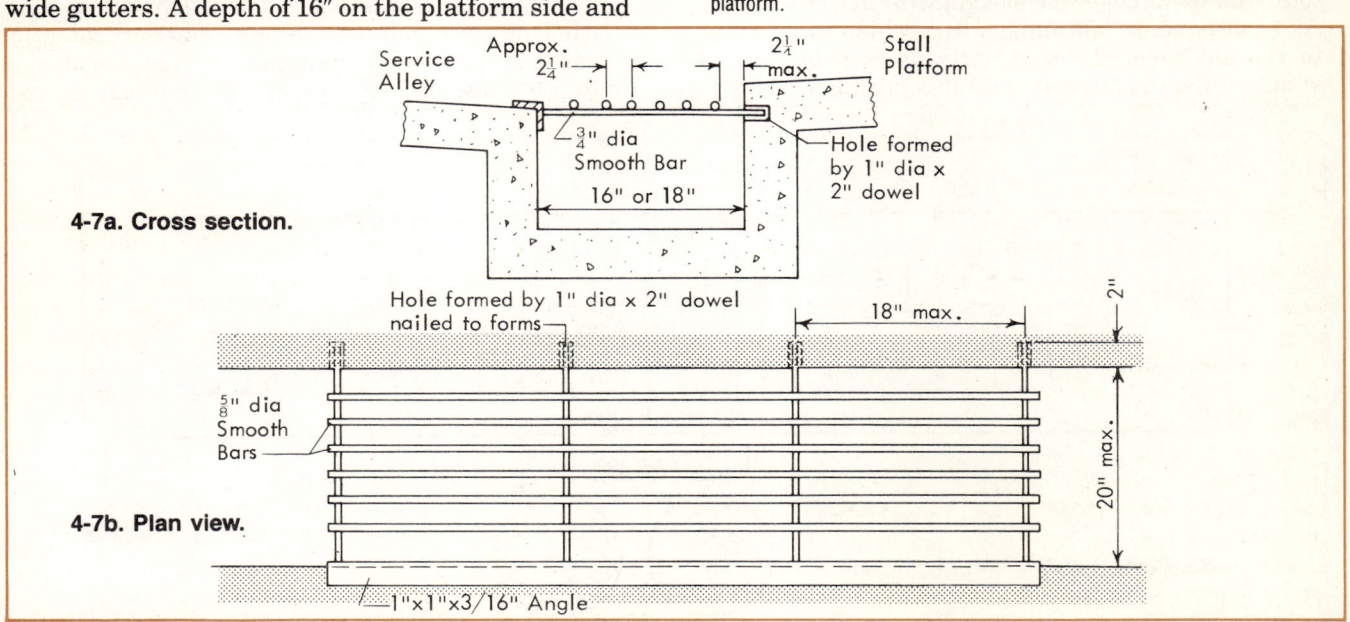

4-7a. Cross section.

4-7b. Plan view.

Fig 4-7. Gutter grates.
Use steel angles on both sides if cow platform is not higher than service alley. Use only 5 bars with a 16" gutter.

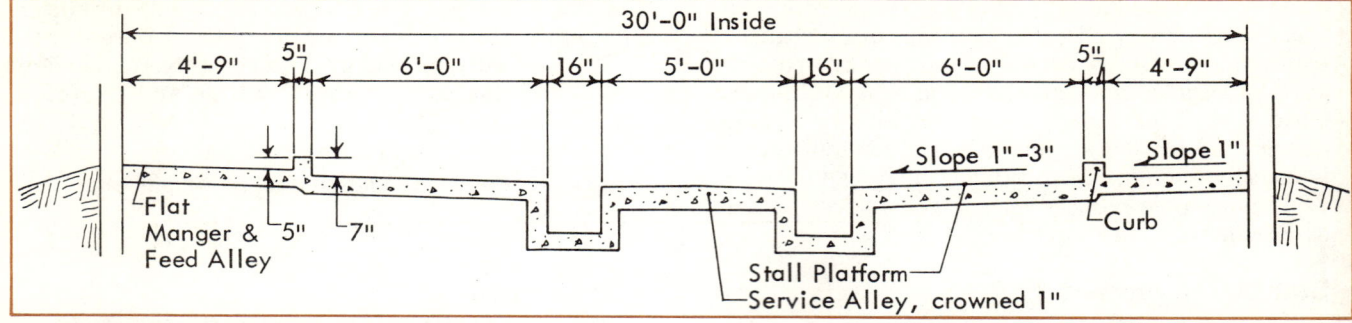

Fig 4-8. Dimensions for remodeling a 30′ stall barn.
Remodel to provide long stalls and still allow using a feed cart.

Barn Remodeling

Generally, it is much easier to remodel a clear-span, single-story barn than a two-story barn with internal posts. Often the mow floor posts and supports of two-story barns are between the stalls and are difficult to change. A new floor, stalls, and waterers may be necessary.

Remodeling is different from new construction because stalls must fit an old building of fixed dimensions. Generally, an old barn can be lengthened for more capacity, but it is impractical to increase width.

Remodeling is usually to provide more comfortable stalls and to make the arrangement more convenient. Make stalls the recommended size, Table 4-2, and gutters at least 16″ wide. Other dimensions depend on the remaining space. Litter alleys can be 5′ wide or less. Make a drawing of all proposed changes before any work is begun. Do not consider remodeling barns less than 30′ wide inside.

When lengthening a barn, install a ventilating system in the addition, but also correct any ventilation problems in the existing building. Old barns generally have insufficient fresh air intake; add more intakes to balance exhaust and intake between the new and old. An exhaust and intake system for the total number of cows may be installed in the addition. As a consequence, animal health problems may occur in the old barn, the new barn, or both because of widely different environmental conditions.

Stall Barn Plan

Fig 4-9a shows a floor plan for a 60-cow stall barn. See Table 4-1 for stall sizes. Alleys are planned for a mechanical silage cart. One or both cross alleys can be eliminated depending on silo placement. Use a 4′-6″ high solid partition on the box stall sides adjacent to cow stalls to reduce exposure to disease organisms. If freshening in box stalls, ventilate from the box stalls to the area occupied by the milking herd. If possible, put freshening pens in a separate area. Slope barn floors about 1″ per 10′ toward the milkhouse so that the milk line height is nearly constant with respect to the floor. Place the milkhouse at the end of the barn for convenient pipeline milker installation.

Free Stall Barns

Free stalls were first used to replace deep manure packs in loose housing systems. Free stalls reduce bedding needs and keep cows cleaner than manure packs. Free stall barns are practical and common for herds of about 60 or more and are usually used for herds of 100 or more.

With properly designed free stalls, most manure is deposited in alleys. Clean solid floor alleys daily with a tractor-mounted blade, mechanical alley scraper, or flushing. Move manure to a cross conveyor, pump, storage area, or spreader. Mechanical scrapers

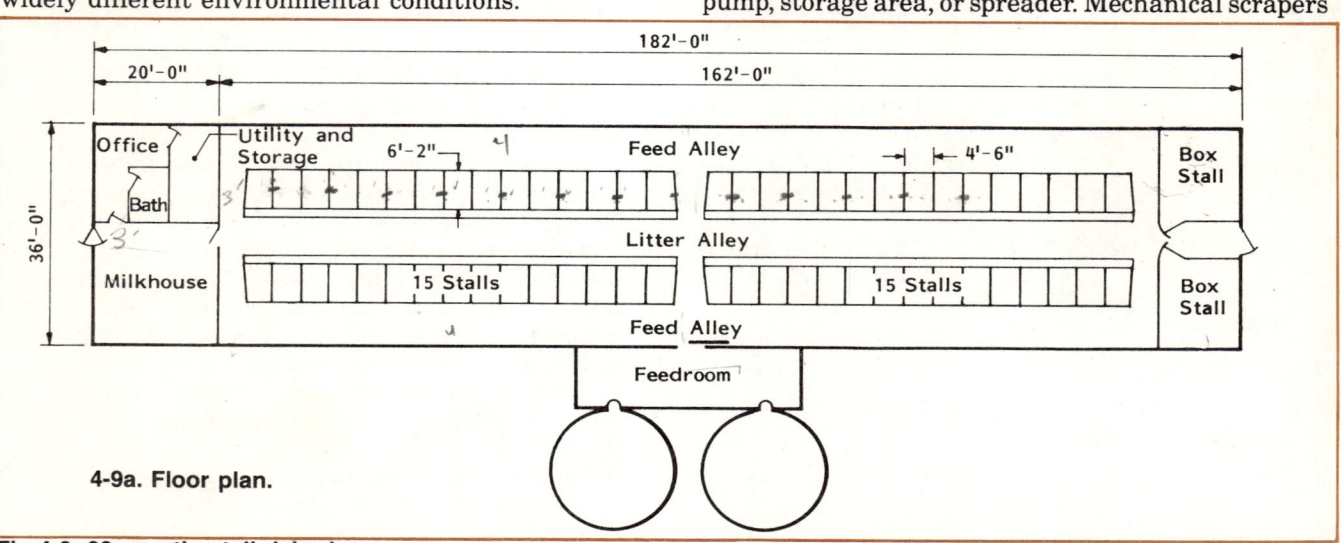

Fig 4-9. 60-cow tie stall dairy barn.

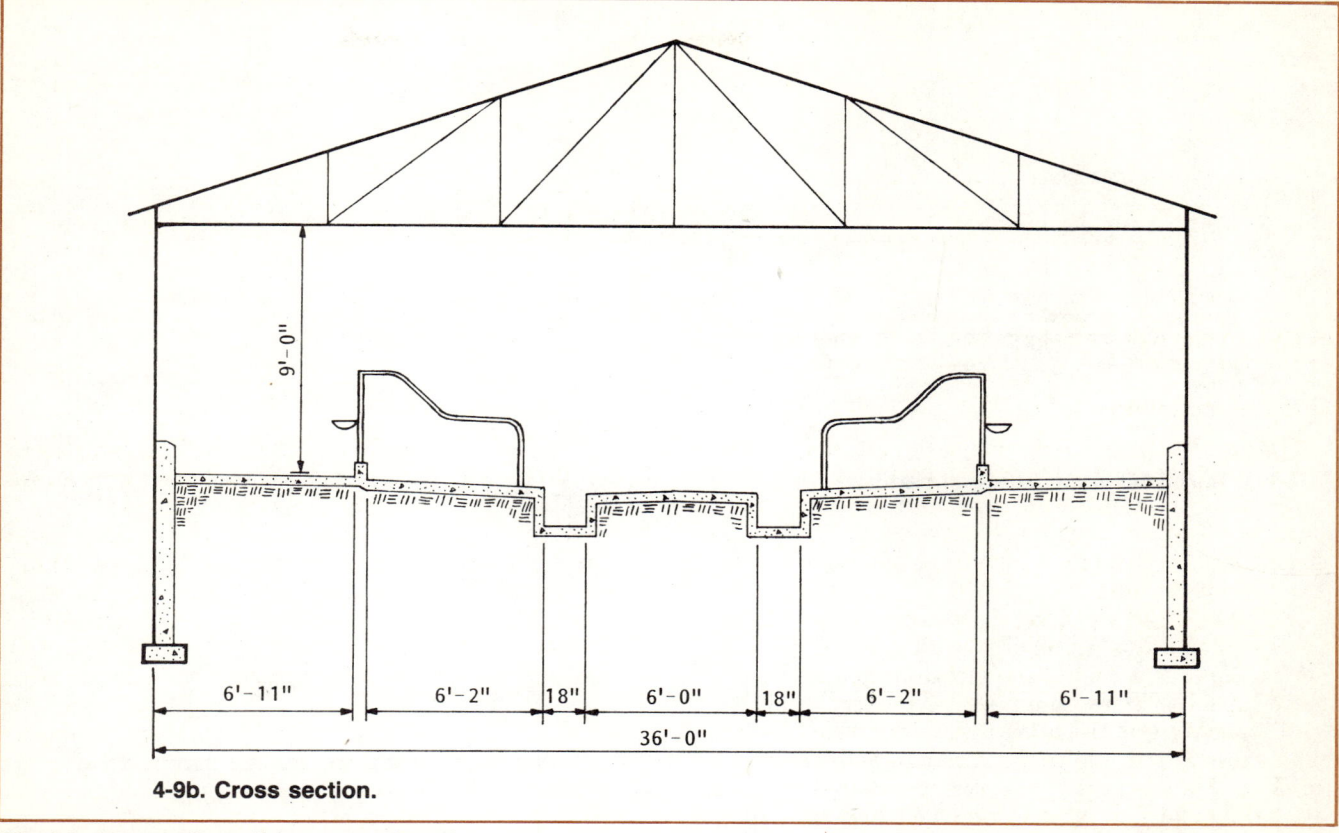

4-9b. Cross section.

Fig 4-9. 60-cow tie stall barn, continued.

and flushing can remove manure while cows are in the stall area.

Slotted alleys eliminate scraping because manure passes through slots into a storage below. See the manure management chapter.

Free stall barns are either "warm" (well insulated and mechanically ventilated) or "cold" (little or no insulation and naturally ventilated). Feeding is usually in the barn or an adjacent feeding area.

Free Stall Design

Make free stalls wide enough for cow comfort but not wide enough for a cow to turn around and drop manure at the front of the stall.

Make stalls long enough so cows lie with their udders far enough forward to prevent injuries, but avoid manure in the stall. See Table 4-3. Provide 4' high side partitions and 5' high front partitions so cows do not hang their heads over. Increase the height if necessary.

The free or alley end of free stall dividers must

withstand pushing by cows. Make partitions open for good air movement. Space rails less than 6" or more than 12" so cows do not get caught in them.

Neck boards across the top stall rail force standing cows back so manure is dropped in the alley. Adjust the neck board forward or backward to keep stalls clean. Cows can get caught and injured under a single neck board. Two or three horizontal rails at the front allow a cow to lie down comfortably, Fig 4-10, and not get caught under the top rail. Do not use cables for neck rails because of injury to cows.

For a comfortable resting area, place bedding even with the curb at the rear and 2"-4" above the curb at the front. Sawdust, chopped straw or corn stover, and sand are common bedding materials. Bedding type is usually determined by available materials and limitations of the manure handling system.

Cows seldom select a hard surface free stall if a soft surface is available. Consider all alternatives before deciding to use rubber mats, concrete, or similar materials. Hard surface free stalls do not reduce the amount of bedding needed. Approximately the same amount of bedding is needed for all free stall surfaces. Slope the free stall the same for all surface and bedding materials.

Bedding boards reduce the formation of holes in dirt or clay free stalls, Fig 4-11. Install the board so the cow does not get her foot under it. The bottom of the bedding board must be embedded in the fill. Bedding boards also reduce udder injuries and aid cows in standing up.

Table 4-3. Free stall dimensions.
Stall width measured center-to-center of 2" pipe dividers. For wider divider dimensions, increase stall width accordingly. Stall length measured from front of stall to alley side of curb.

Cows	Width	Length
1,000 lb	3'-6"	6'-10"
1,200 lb	3'-9"	7'-0"
1,400 lb	4'-0"	7'-0"
1,600 lb	4'-0"	7'-6"

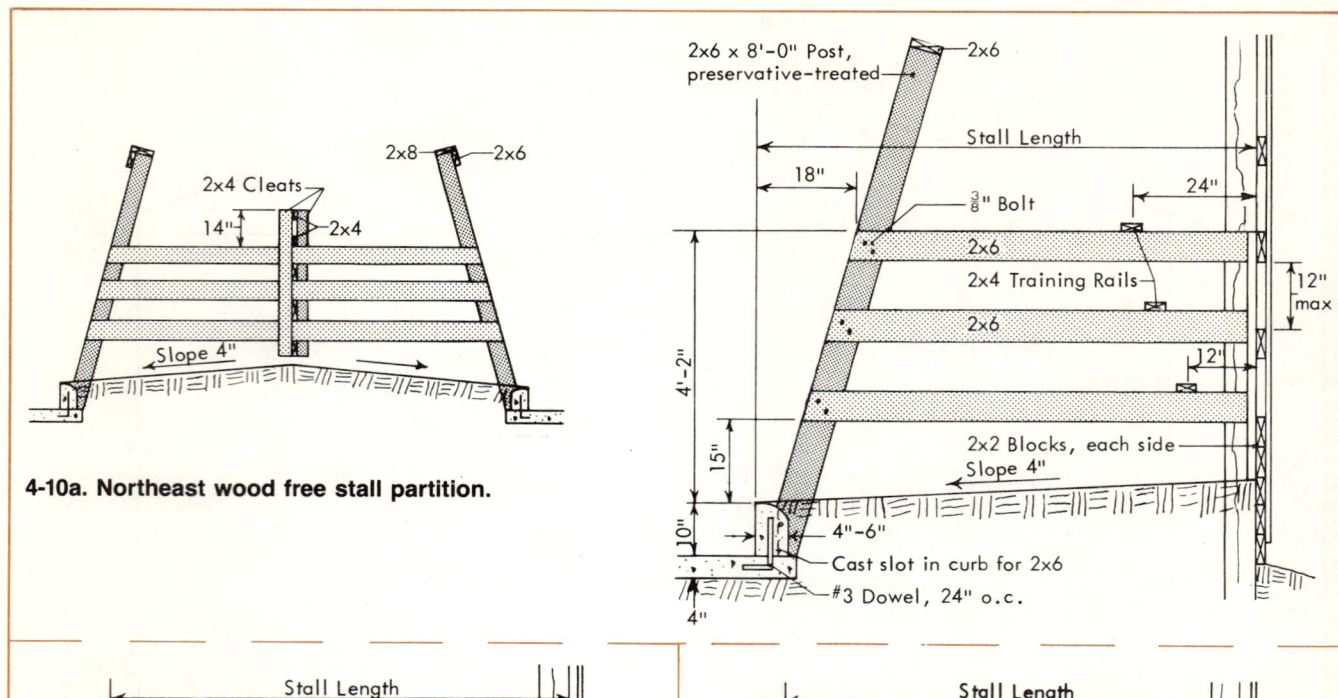

4-10a. Northeast wood free stall partition.

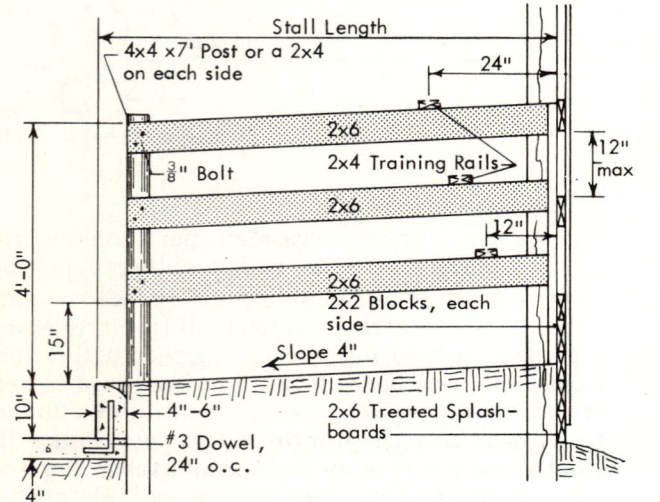

4-10b. MWPS wood free stall partition.
Set rails parallel to the stall floor for constant partition height. Facing free stalls can have either a center supporting partition and sloped rails or horizontal rails and a suspended stall front like Fig 4-10a.

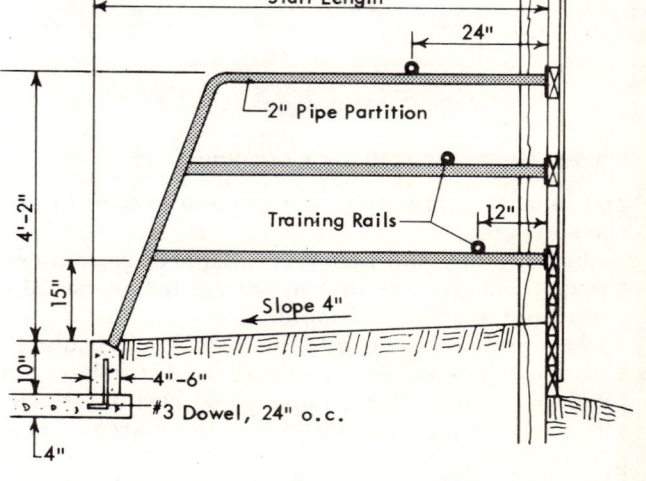

4-10c. Pipe free stall partition.
A fourth rail may be added to prevent cows from sticking their heads through the partition.

4-10d. Suspended free stall partition.
Securely anchor the support brackets and brace the partition with the neck rail. There is no alley post, so install partitions after concrete work is complete. The partitions are 1' shorter than the stall to reduce damage, can be easily replaced or used in remodeling, and allow maintenance of the stall platform with a tractor-mounted blade.

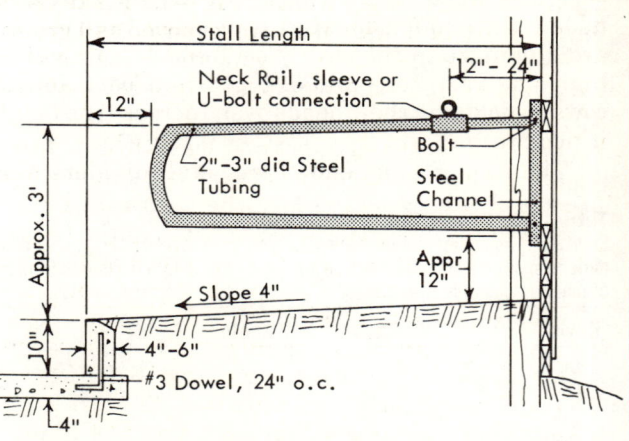

Fig 4-10. Typical free stall construction.
Install a 2x10 bottom partition board, except in Fig 4-10d, if cows work fill and bedding from stall to stall creating deep holes.

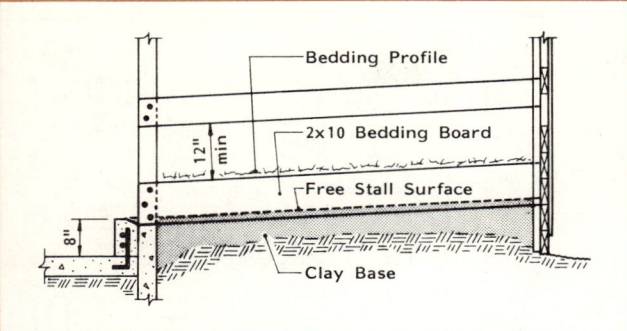

Fig 4-11. Bedding board.

Barn Design

One stall is usually provided for each cow. For certain housing systems and management practices, more cows can be kept in the barn than there are free stalls. Cows do not always use the same stall or all lie down at the same time. Some cows may not use the free stalls but lie in the alleys, particularly in a herd accustomed to stall or loose housing. Remove these cows from the herd.

Detecting cows in heat is similar to other housing types. Although heat periods are obvious, they often occur at night and outward signs are not detected. Contact between cows and producer is reduced with free stall barns.

Finish floors to reduce slipping, avoid injuries, and decrease mastitis and "silent" heats. See MWPS AED-19, *Slip Resistant Concrete Floors*.

Space wall girts 12" or less to protect the siding. Anchor splashboards securely outside the posts, or install them inside to prevent the fill from pushing the wall out. If cows have access to the outside of the building, protect the bottom 4' of the wall with wood siding, treated plywood, or splashboards.

Use an 8" curb with slotted alleys.

Barn Layout

Free stall barns commonly have 2, 3, or 4 alleys, which are usually for both feed and stall access. Alley width depends on whether it is for access to stalls, feeding, or both. Avoid dead-end alleys. Barn width is directly related to alley widths, Fig 4-12. Make cross alleys 4' wide, or 8' wide if cows are moved as a group. With mechanical scrapers, use narrower stall access alleys. Slope solid floor alleys to storage areas or cross conveyors.

Free stall arrangements for at least three production groups are common. Separate high and low producers and dry cows, and feed them accordingly.

In solid floor barns, use 8"-12" high curbs to protect walls, free stalls, bunks, and waterers from damage during scraping. Elevate fences, waterers, and cow cross alleys from manure collection areas.

Provide at least 1 cup waterer or 2' of tank perimeter per 20 cows. Limit water depth to 6"-8" for fresher water and less debris accumulation. At least 2 waterer locations are needed for each group of cows.

If cows are fed outside the barn, locate waterers for outdoor access. Slope the ground surface away from waterers for better drainage.

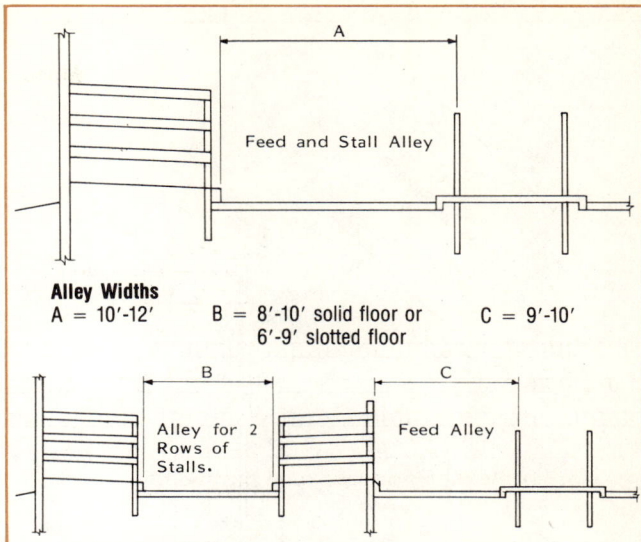

Fig 4-12. Alley types.

Alley Widths
A = 10'-12'   B = 8'-10' solid floor or 6'-9' slotted floor   C = 9'-10'

Barn Plan Descriptions

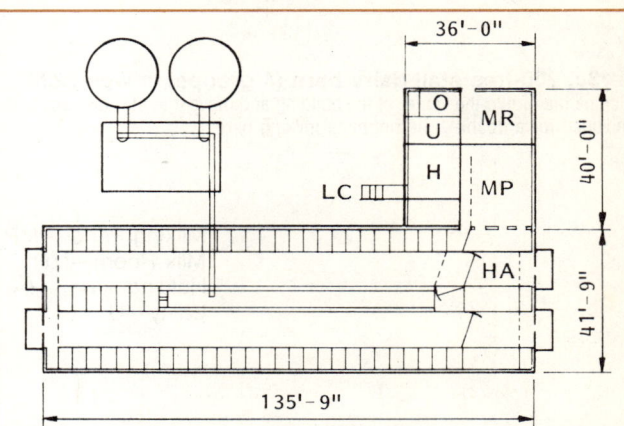

4-13a. 62-free stall dairy barn (2 groups), mwps-72352.
Scrape manure to one end of the building for removal. Space is included for a double-4 herringbone milking parlor. To expand the barn to 124 free stalls, locate the feed center across the barn from the milking center and build a similar free stall area to the right and a second return alley.

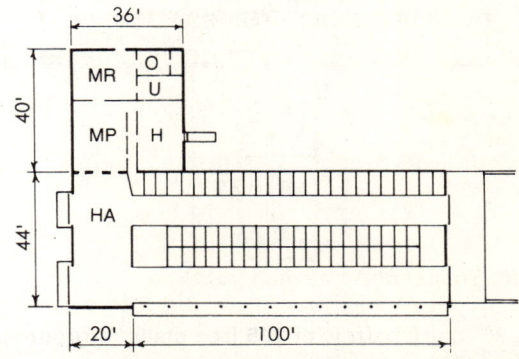

4-13b. 64-free stall dairy barn (2 groups), mwps-72353.
The fenceline bunk is protected by a roof overhang for outside filling. Scrape manure to one end of the building for removal. Space is included for a double-4 herringbone milking parlor. To expand the barn to 128 free stalls, build a similar free stall area to the left and a second return alley.

Fig 4-13. MWPS cold dairy barn plans.
These pole buildings are naturally ventilated. Plan building layout for adequate ventilation. The herringbone milking centers include a milking parlor, milkhouse, utility, shower, and office.

4.8

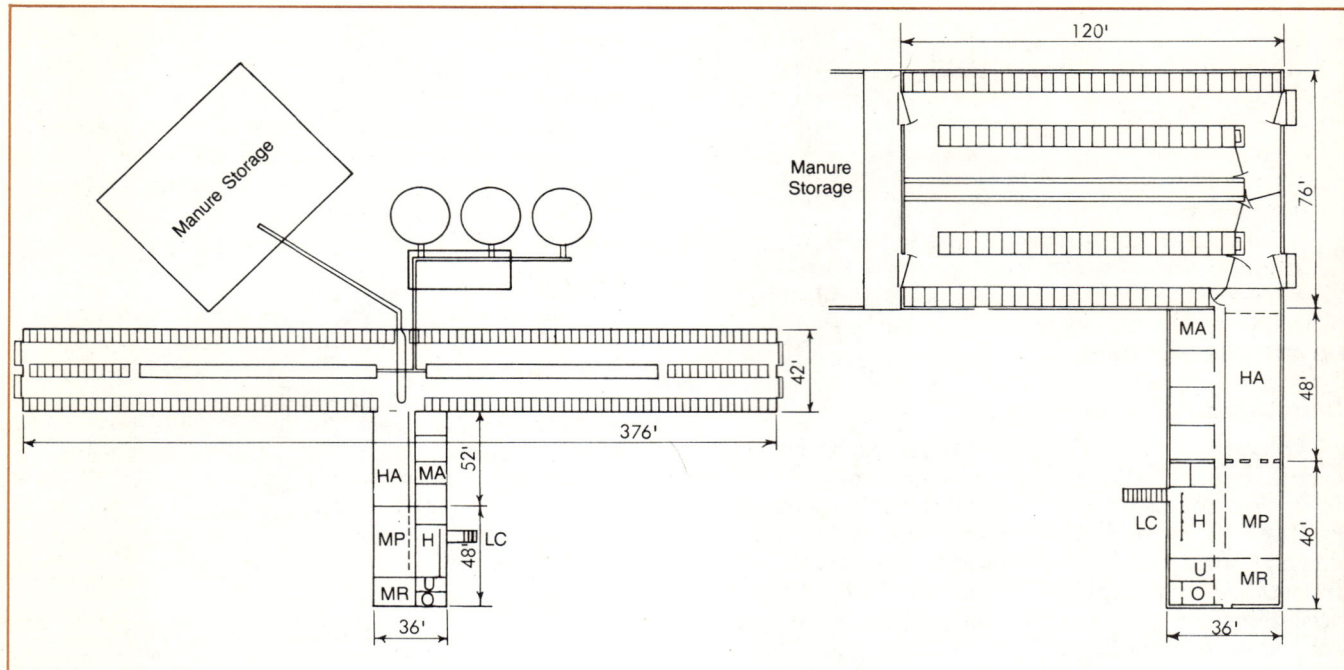

4-13c. 200-free stall dairy barn (4 groups), mwps-72354.
Scrape manure to the center of the building and move it to storage. Space is included for a double-8 herringbone milking parlor.

4-13e. 76′ dairy barn with 100 free stalls (2 groups), mwps-72356.
Space is included for a double-6 herringbone milking parlor. Expand the barn by building a similar free stall area to the right.

Milking Parlor—MP Office—O
Milk Room—MR Loading Chute—LC
Holding Area—HA Hospital—H
Utility—U Maternity—MA

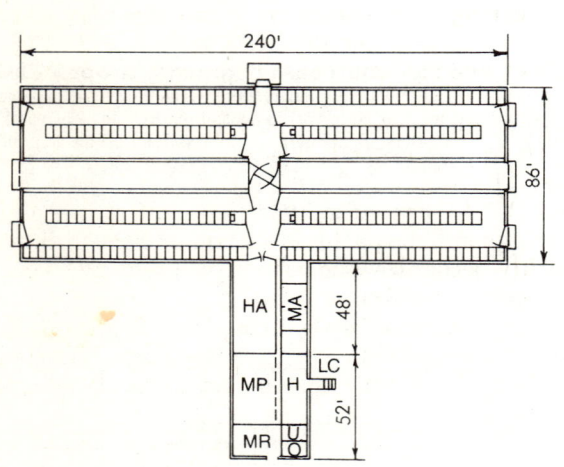

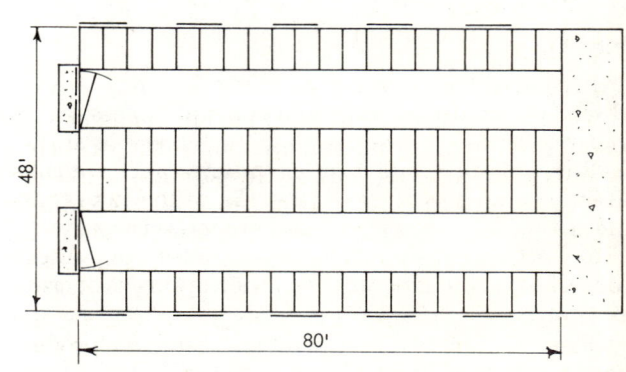

4-13d. 86′ dairy barn with 206 free stalls (4 groups), mwps-72355.
Manure is tractor-scraped to the center of the building and moved to a storage area opposite the milking center. Space is included for a double-8 herringbone milking parlor.

4-13f. 80-cow free stall dairy barn, resting only, mwps-72361.
Cows are fed in an outside lot and milked in a separate building.

Fig 4-13. MWPS cold dairy barn plans, continued.

5. MILKING CENTER

A milking center for free stall housing includes a milking parlor, holding area, utility room, milk room, lounge area, office, and treatment area. A stall barn milking center has a utility room, milk room, lounge area, and office. Milking in a stall barn is done in the stalls using bucket units or an around-the-barn pipeline.

The U.S. Public Health Grade A Pasteurized Milk Ordinance requires approval of a plan for each farm before construction or remodeling begins. Contact your milk buyer about approval procedures before building. Approval agencies usually require an overall plan drawn to scale and dimensioned, showing lighting, ventilation, insulation, waste disposal, and water supply.

Milking center dimensions depend on milking parlor type, cow size, milk tank size and type, compressor location, and type of washing and milk handling equipment. Coordinate construction with your building contractor and equipment supplier.

Milking Parlors

Milking parlor type is the first consideration in designing milking centers. Parlor type influences:
- Milking center size, layout, and location.
- Cow traffic patterns.
- Milking routine.
- Amount of mechanization.

Parlor size depends on:
- Initial mechanization.
- Plans for future improvements.
- Number of cows milked.
- Available labor and capital.
- Milking time available.
- Milk production level.

Parlor selection also depends on:
- Herd expansion.
- Initial investment.
- Annual costs.
- Personal preference.

When selecting a milking parlor, consider total chore time which includes parlor setup, milking, group changing, and cleanup. To estimate total chore time, assume 20 min for parlor setup, 30 to 45 min for parlor cleanup, and 15 min per 100 cows for changing groups. Times vary with parlor size and layout. Estimate milking time from capacities in Table 5-1.

Capacities are useful for comparing and estimating milking time, but overall labor efficiency is important to selecting a milking parlor. Parlor type affects the amount of time spent milking. The percentage of chore time spent milking varies from one dairy to another. With small dairies, actual milking time is a smaller part of the total chore time than in larger dairies. So investment in a larger parlor and more mechanization to improve capacities in a small dairy results in less improvement in labor efficiency.

Evaluate the benefits and cost per head of improvements to the milking parlor. The least cost parlor system may not always be the best alternative when available labor, milking time, and expansion are considered.

Parlor Types

Flat barn

In the Midwest and Northeast, 6- to 10-stall flat barns have been used as an alternative to more expensive elevated parlors. A flat barn can serve until an elevated parlor can be afforded. Put stalls in a room the size of the future elevated parlor.

Side-opening

Side-opening parlor layouts are usually double-2, -3, or -4 and for herds of less than 250 cows, Fig 5-1. The advantage of side-opening parlors is more individual cow attention, but there is a greater distance between udders compared with the herringbone. This distance becomes important when the parlor is mechanized, and more cows are needed in the parlor to keep the milker and the equipment busy. Pit length increases by 8'-10' for each pair of milking stalls.

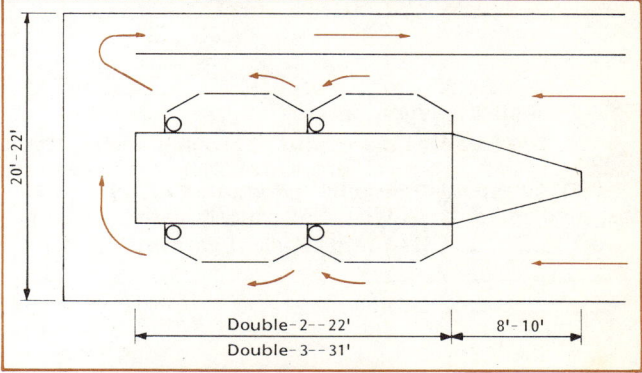

Fig 5-1. Double-2 side-opening parlor.

Herringbone

Herringbones are the most common elevated parlor. They range from double-4 to double-10, Fig 5-2. Cow movement improves when cows are handled in groups compared with being handled individually in side-opening and rotary parlors.

Herringbone stalls are adaptable to mechanization. The walking distance between udders varies from 36"-48" depending on the manufacturer. With a herringbone, the parlor is shorter and the milker can quickly recognize and attend to problems, such as accidental machine drop-off.

Larger herringbones do not increase cow milking capacity or quality because cows are handled in batches. A slow-milking cow can seriously effect cow throughput.

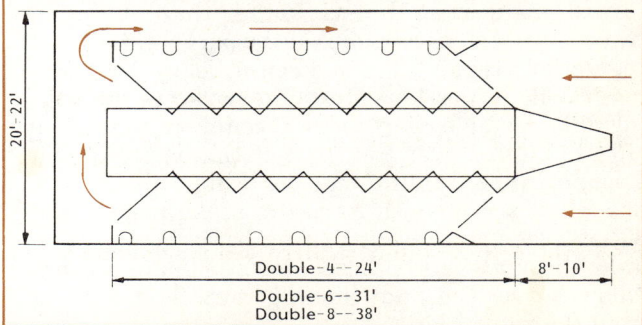

Fig 5-2. Double-8 herringbone parlor.

Polygon

The polygon parlor, Fig 5-3, combines some of the advantages of herringbone and side-opening parlors. Distance between udders is minimized, and with fewer cows per side, fewer cows are held up by a slow-milking cow. Usually, polygons have 4 sides with 6 cows per side. The number of cows per side (4, 5, 6, 8, or 10) depends on the operator's routine, degree of mechanization, and management.

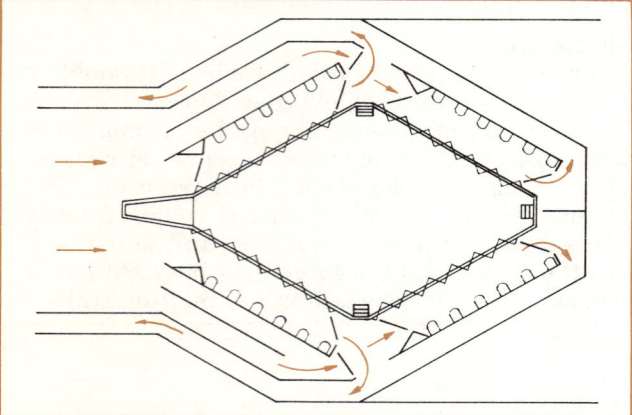

Fig 5-3. 24-stall polygon parlor.

Trigon

A trigon is a three-sided polygon, Fig 5-4. It was designed for dairies with 250 to 500 cows. Trigons have 12-, 16-, 18-, 22-, and 24-stall capacities.

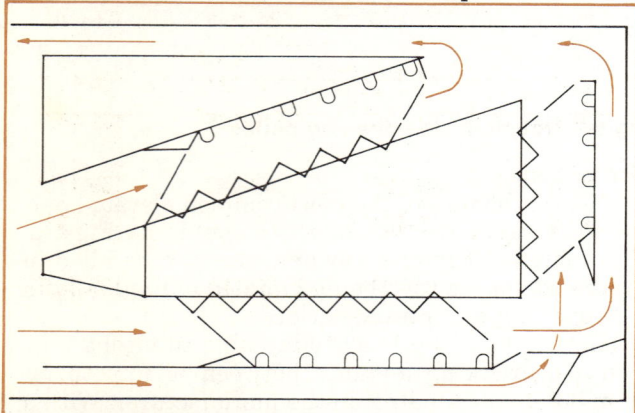

Fig 5-4. 16-stall trigon parlor.

Parlor Layout

A straight entrance from the holding pen into the milking parlor and a straight exit from the parlor is preferred. Turns at the entrance can slow cow movement, interrupting the operator. If cows must be turned, make it at the exit rather than at the entrance. Avoid steps or ramps at the parlor entrance. If regulations require a step, keep it below 8″ high.

Provide return lanes from the parlor to the housing unit. A single return lane is common—one group of cows crosses over the front of the parlor to exit. Return lanes outside the building can be wide enough to scrape with a tractor. Make inside lanes 33″-36″ wide to prevent exiting cows from turning around. Lanes can be hosed down or hand-scraped. Allow for sorting and catching cows for treatment from the return lane.

Parlor Mechanization

Milking parlor mechanization depends on parlor size, available labor, initial investment, and personal preference. Increase parlor size as mechanization is added to better use both labor and equipment. Automatic detachers are usually standard equipment in most large parlors. Other mechanization to consider includes crowd gates, power-operated gates, feedgates, stimulating sprays, and wash and prep stalls.

Holding Area

Holding pens confine cows ready for milking. Cover the area to protect animals from the weather. Use a crowding gate to encourage cows to enter the parlor.

Provide 15 ft^2/cow in the holding area. Consider parlor capacity when planning the holding area. Hold cows for no more than 2 hr at each milking (1 hr in hot climates or with three times daily milking). For parlors that continue milking while changing groups, increase the holding area by 25% to allow for overlap.

With confinement housing, the holding area is usually connected to the barn and all traffic areas are covered. Part of a resting or feeding alley can be the holding area in some arrangements, but use of a crowd gate, future expansion, and division of the herd into production groups is limited. Alley scraping can also be difficult.

When milking two or more groups of cows, a holding area separate from the housing unit is preferred. Cow movement patterns, feeding coordination, manure removal, and milking are simplified.

With an enclosed holding area, replace the wall between the holding area and milking parlor with an overhead door. Movement into the parlor is improved when cows can see parlor activity while waiting in the holding area. If the parlor opens directly into a cold barn, do not remove the wall because the parlor can be very cold in the winter.

Except in severe climates, the holding area need not be insulated. Cows brought in for milking warm the holding area so that opening the overhead door does not lower the parlor temperature except during very cold weather.

If the wall is removed and no door is used, insulate the holding area to conserve heat in the winter. Use heat from the bulk tank refrigeration system to heat the parlor and holding area.

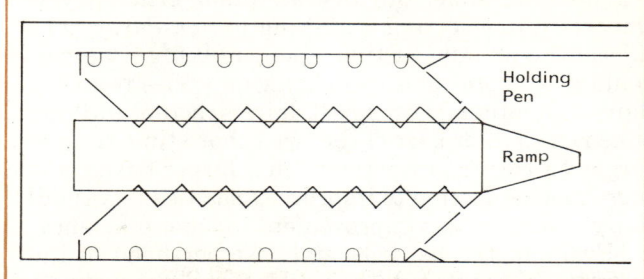

Fig 5-5. Parlor with ramp.
A ramp makes it easier to get behind balky cows without chasing them away from the entrance.

Table 5-1. Milking parlor cow capacities.
D = double, e.g. D-2 = 2 rows of 2 stalls. Low value of each range is for cows producing about 60 lb/day; high value is for cows producing about 38 lb/day. Values are steady-state capacities—parlor setup and cleanup, and group changing not included. Assume parlors have power-operated entrance and exit gates. Superscripts ([1], [2], [3]) denote number of operators milking.

5-1a. Side-opening and herringbone parlors.

Mechanization	Side-opening D-2	Side-opening D-3	Herringbone D-4	Herringbone D-6	Herringbone D-8	Herringbone D-10
	- cows/hour -					
None	25-35[1]	50-63[2]	29-42[1]	50-66[2]	64-80[2]	80-89[2]
Crowd gate	28-38[1]	52-65[2]	34-47[1]	55-71[2]	69-87[2]	88-97[2]
Prep stalls	34-44[1]	52-65[2]				
Crowd gate and prep stalls	38-48[1]	56-69[2]				
Crowd gate and feedgates			37-47[1]	58-74[2]	72-90[2]	92-101[2]
Detachers			33-46[1]	49-65[1]	60-78[1]	72-81[1]
Detachers and crowd gate	36-46[1]	44-57[1]	37-50[1]	54-70[1]	68-84[1]	79-88[1]
Detachers, prep stalls, and crowd gate	40-50[1]	50-63[1]				
Detachers, crowd gate, and feedgates			39-52[1]	57-73[1]	70-88[1]	83-92[1]

5-1b. Trigon and polygon parlors.

Mechanization	Trigon 12-stall	Trigon 16-stall	Trigon 18-stall	Polygon 16-stall	Polygon 20-stall	Polygon 24-stall	Polygon 32-stall
	- cows/hour -						
None	53-74[2]	68-89[2]	76-95[2]	71-97[2]	86-112[2]	101-127[2]	121-157[3]
Crowd gate	59-79[2]	74-96[2]	83-103[2]	78-104[2]	94-120[2]	110-136[2]	131-167[3]
Crowd gate and feedgates	62-82[2]	78-100[2]	87-107[2]	83-109[2]	98-126[2]	117-143[2]	139-175[3]
Detachers	50-70[1]	63-85[1]	67-87[1]	68-94[1]	75-101[1]	79-107[1]	117-153[2]
Detachers and crowd gate	56-76[1]	71-92[1]	75-94[1]	76-102[1]	83-109[1]	90-116[1]	129-165[2]
Detachers, crowd gate, and feedgates	59-78[1]	73-95[1]	78-98[1]	81-107[1]	89-115[1]	96-122[1]	137-173[2]

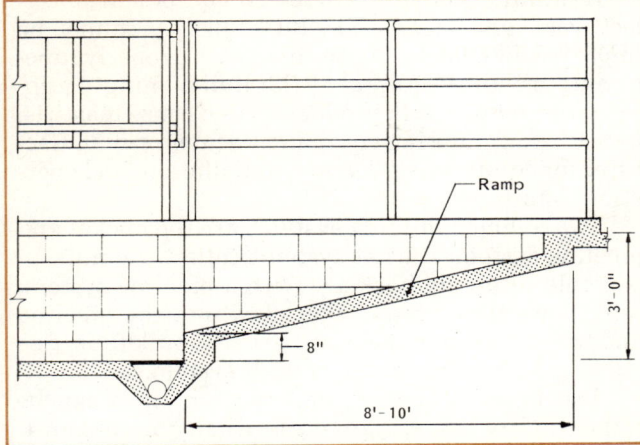

Fig 5-6. Interior side view of ramp and pit.
The ramp slopes up from a step at the back of the pit through a distance of 8'-10' to a step into the holding pen. A gutter with a grate drains the pit and ramp. Provide good footing for the operator.

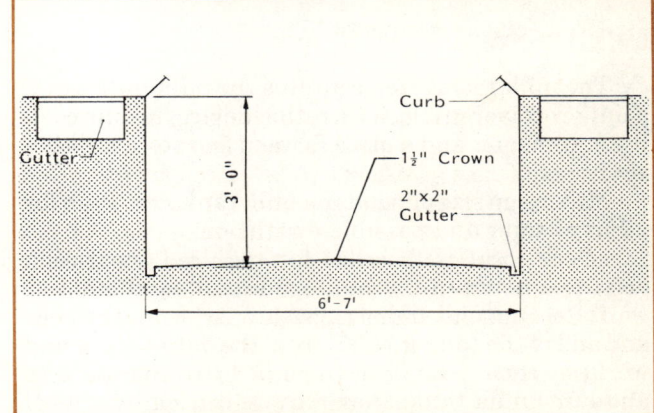

Fig 5-7. Pit cross section.
A 1½" crown reduces strain on the operator's legs and directs water to pit walls. A 2"x2" gutter along each pit wall carries water to the gutter at the end of the pit. Slope the pit floor and gutters toward the end of the pit. Use a rough broom finish on the pit floor for operator safety.

Utility Room

Provide storage for milking and processing equipment. Separate equipment from walls by at least 3' for maintenance.

Install good lighting, a floor drain, and good ventilation. Ventilation is essential to cool compressors and vacuum pumps. Design a ventilating system to use heat from the compressor to warm the rest of the milking center in the winter. See the building environment chapter for recommendations.

Electrical service is required for vacuum pumps, milk cooling equipment, water heaters, ventilating fans, furnace, grain meters, and electric fence controller. Locate the electrical entrance panel on an interior wall of the utility room to reduce condensation and corrosion of electrical contacts. For safety and to meet National Electrical Code requirements, do not locate equipment within 3½' in front of, above, or below the panel. Do not locate other utility lines such as water within 3½' of the entrance panel.

Table 5-2. Utility room space.

Item	Area, ft²
Milk vacuum pump	6-9
Compressor	8-10
Water heater	4-6
Furnace	3-5
Storage	3-5
Work alleys	20-30
Refrigerator	6-9
Desk	4-12

Storage Room

Provide a separate room for storage of cleaning compounds, medical supplies, replacement milking system rubber components, and similar products. Separate this space from the utility room to reduce deterioration of rubber products from high temperatures, light, and ozone associated with motor operation. An interior, windowless space with ventilation to control temperature rises is recommended. Provide this area with good lighting, a floor drain, and two refrigerators—one for medical products and a second for replacement rubber products.

Milk Room

The milk room often contains the bulk milk tank, a milk receiver group, a filtration device, in-line cooling equipment, and a place to wash and store milking equipment.

Milk room size depends on bulk tank size. Plan for a larger tank and possible expansion.

Check local milk codes for required separation distances between the bulk tank and equipment or wall. Recommended distances are 24" from the rear and end of the tank and 36" from the outlet valve and working ends. Gravity-type milk filtration systems and large bulk tanks may require ceiling heights of 10' or more.

Many larger bulk tanks are designed so that a major portion of the tank extends through one wall (bulkheaded) to the outside or to an adjacent utility room, Fig 5-8. This reduces milk room size and the cost of building materials.

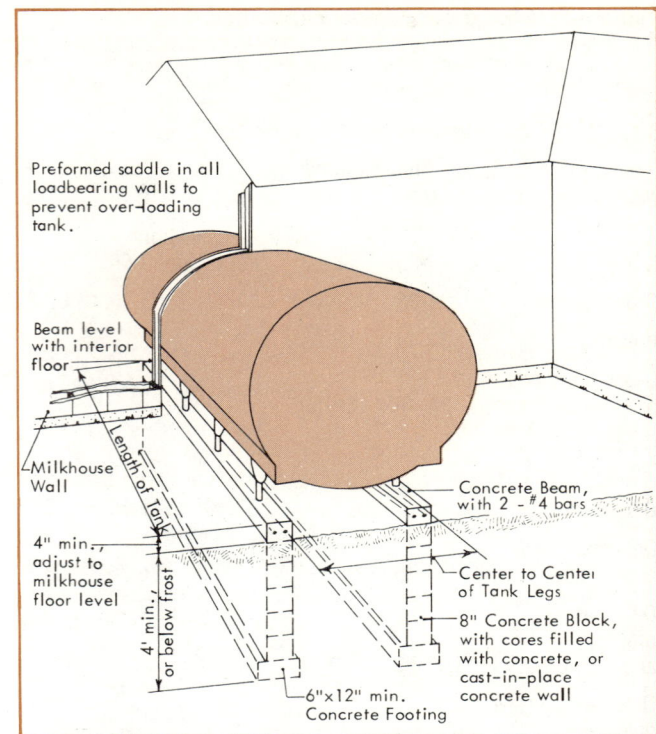

Fig 5-8. Masonry footing for bulkheaded milk tank.

Toilet and Office

Provide at least a toilet and sink in the milking center. Consider a shower, lockers, and a resting area for hired workers. Separate facilities for men and women may be required by local regulations. For sanitary reasons, do not open the toilet room door into the milk room. See the manure management chapter for waste handling methods.

A milking center office is needed for keeping herd health and production records. It may also be the main farm office. Protect computer-based feeding and record systems from dirt and moisture.

Milking Equipment

A milking system includes all equipment to collect, cool, and store milk. Milk quality cannot be improved, but can be maintained with properly functioning milking equipment. Malfunctioning equipment can reduce milk production and cause mastitis. Consult your milking equipment dealer or manufacturer for equipment selection, installation, and operation details.

Two common milking systems are the bucket and pipeline system. Bucket systems collect the milk in portable receiving buckets. With a pipeline system, milk flows through a milker claw into the pipeline and travels through the pipeline to the bulk tank. See Table 5-3 for components of each system.

The location of milk and vacuum lines can be either high or low. A high line is usually more than 4' above the cow platform and a low line is below the cow platform. Low milk lines are preferred so that milk flows down to the pipeline requiring a lower vacuum level (11"-13" Hg). With a high line, the milk must be lifted up to the milk line requiring a higher vacuum

Table 5-3. Milking system components.

Component	Bucket	Pipeline
Vacuum pump(s)	X	X
Balance tank/header	X	X
Vacuum controller(s)	X	X
Vacuum line and stallcocks	X	
Pulsation line and stallcocks		X
Vacuum supply line to header/trap		X
Bucket unit with claw, shells and liners	X	
Pulsator—electric, master or pneumatic	X	X
Milk pipeline		X
Weigh jars (optional)		X
Receiver, pump, trap		X
Strainer or milkveyor plus filter	X	
Filter system		X

level (13"-15" Hg). Low lines reduce vacuum fluctuations which can cause damage to the udder.

Install milk lines and vacuum lines in a complete loop with a double inlet at the receiver to reduce vacuum fluctuations. Dead end vacuum lines cause increased vacuum fluctuations. Slope milk lines 1" - 1½"/10' towards the receiver to improve milk flow. Slopes less than 1"/10' increase the chance of milk line flooding and vacuum fluctuations. Higher slopes cause more milk foaming and increased rancidity. Milk line size depends on the number of milking units per slope. See Table 5-4.

Vacuum pump and vacuum controller size depend on manufacturer's specifications and the total air flowrate requirement of each milking system component. Size the vacuum controller to have a capacity equal to or greater than the vacuum pump. Provide at least 8 cfm (American Standard) per milking unit at the vacuum pump and at least 3 cfm per unit at the receiver. Lower vacuum pump capacities could cause increased inflation squawking, machine fall-off, and vacuum fluctuations. Provide a reserve pump capacity of at least 30 cfm. For help determining your vacuum pump and airflow needs, contact your milking equipment dealer.

Table 5-4. Milk and vacuum lines.

Milk line diameter
2 units/slope	1½"
4 units/slope	2"
6 units/slope	2½"
9 units/slope	3"

Milk line slope
Stainless steel	1"/10'
Glass	1½"/10'

Vacuum line diameter
2 to 4 units	1¼"-2"
5 to 7 units	1½"-2½"
8 to 12 units	2"-3"

Milk Cooling

Some milk cooling devices are:
- Conventional "compressor/condensor" bulk tank cooler. Size compressors according to milk flow per hour entering the tank.
- Combination of heat exchanger and bulk tank cooler. Plate coolers and tube coolers are between the receiver jar and the bulk tank. Well water runs countercurrent to milk flow through the heat exchanger. Milk temperature decrease depends on heat exchanger surface area, water temperature, and water and milk flowrates.
- Heat exchanger plus ice bank. Ice water is circulated through the heat exchanger in place of well water. Milk enters the bulk tank at about 36 F. The bulk tank maintains milk temperature and does not cool the milk.

Milk Heat Recovery

Heat recovery equipment is practical for almost all dairies. It becomes less economical with smaller dairies. Consider the advantages of heat recovery equipment to your operation before making a selection.

Heat exchangers are available to heat air and water with refrigeration systems. Heat from an air-cooled condensing unit can heat the milking center.

A desuperheater heat exchanger removes heat from the refrigerant gas after it leaves the compressor. Concentric tubes carry water and refrigerant gas in a countercurrent pattern. Depending on compressor hp and operating time, water can be heated up to 180 F.

Water-cooled condensors can heat water. Because water is the condensing medium, the condensing temperature typically limits the heated water temperature to 110 F, but in some units, temperatures can reach 180 F.

Water Supply

Wash and sanitize udders with 115 F water in the milking parlor. Wash milk handling equipment, etc., with 165 F water in the milk room. Two separate water heaters are recommended, but 105 F water can be obtained from 165 F water through a mixing faucet. Insulate all hot water pipes with preformed insulation.

Pipe hot and cold water to all wash and rinse vats in the milk room. Provide a mixing faucet with a hose and nozzle for cleaning the bulk tank, floor, and walls. Dispense acidified water and/or sanitizer through this hose and nozzle to improve sanitation and milk quality. CIP (clean-in-place) cycle rinse water can be retained in a storage tank and used to wash the floor.

For some installations, a hot water booster heater is needed. A calrod heater can be used in an open wash tank or installed in a stainless steel jacket as part of the hot water line. These heaters are controlled by the automatic washing equipment and a thermostat.

Table 5-5. Milking center water heater sizes.

Parlor size	Washing cows 115 F	CIP water 165 F
Double-4	50 gal	80 gal
Double-8 or milking barn	80 gal	120 gal

Milking Center Construction

Material and construction requirements vary with milking center operating conditions. Consider operator use, animal contact, sanitation, fire risk, temperature, and moisture levels. Proper construction and materials can reduce maintenance and improve operation. Also base selection on initial cost and compatibility with existing structures.

Wall Construction

Most milking center walls are stud or post frame construction. Use batt insulation in the wall cavity with a vapor barrier and protective interior liner.

When an insulated wall is not needed, masonry block walls are durable and economical. Nonbearing walls can be 4" thick and can resist animal impact. Waterproof epoxy paint or tile-faced block on the inside resists moisture penetration and is easily cleaned. Stud frame, post frame, or masonry construction can be used separately or in combination.

Inside Lining Material

Milkroom and milking parlor activities require wall linings that are easily cleaned, resistant to cleaning detergents and acids, impervious to water, light-colored, and durable. Select lining material for cleanability, durability, and initial cost.

Fire rated fiberglass reinforced plastic panels, supported by plywood, are one of the best alternatives. The high initial cost is offset by cleanability, durability, and low maintenance. Masonry walls sealed and painted with a good epoxy paint are another alternative when insulation is not needed. Other materials are asbestos cement board, exterior plywood, glazed tile, and glazed block. Joints should be carefully finished and sealed. Aluminum and steel sheets are suitable for ceilings.

In other milking center rooms, plywood, chipboard, or similar materials can be used. Seal all materials to keep moisture out and improve cleanability. Tightly seal all joints to protect building materials.

In animal traffic areas, a rub rail 36" above the floor protects wall surfaces. A rub rail can be a 2" pipe spaced about 2" from the wall.

Insulation

Insulate rooms that will be kept warm in the winter. Provide the same insulation levels that are recommended for warm barns. See the insulation section. Use 4 to 6 mil polyethylene vapor barriers to protect insulation. Be sure there are no breaks or holes in the vapor barrier.

Floors

Concrete floors are durable, cleanable, waterproof, and safe. A compressive strength of at least 4,000 psi is needed for the milking center floor. Use concrete with a maximum aggregate size of ¾" with a minimum water-cement ratio of 0.49 lb water/lb cement. High quality concrete is more resistant to deterioration from milk and cleaning compounds. When placing concrete do not overwork the surface. Imbedding a piece of slate or tile under milking equipment and bulk tank drains can reduce chemical damage. Foremilk onto a black tile in the floor of each milking parlor stall to check for clinical mastitis.

Where a nonskid surface is needed for cow and human traffic, trowel carborundum or aluminum oxide chips (1 lb/4 ft^2) into the surface. A stiff broom finish is a common alternative. Groove the concrete in the holding areas. For more information about placing and curing concrete and floor surfaces, see the concrete chapter.

Install curbs at doorways to prevent water from flowing between rooms. A curb on the parlor platform reduces splashing. Use preformed stainless or galvanized steel curbing.

Drains

Use deep-water-seal trap drains, Fig 5-9. These drains have high flowrates and provide continuous trapping, preventing gas backflow. Install sewers, drains, fittings, and fixtures according to your state plumbing code. Use materials approved in the code. Cast iron sewer piping with approved seals is recommended. Use at least 4" drain lines.

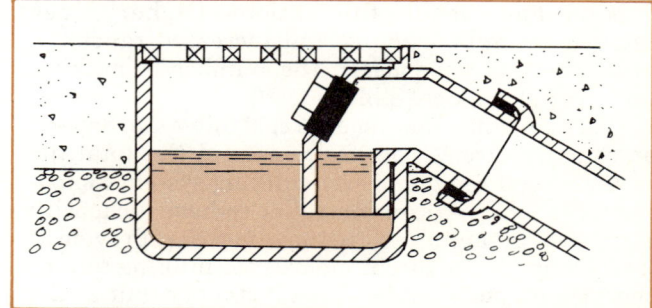

Fig 5-9. Deep-water-seal drain.

Avoid locating drains in the middle of large floors in milk rooms and milking parlors. Locate drains in gutters or room corners to improve drainage. Recess drains ½" below the floor or gutter surface. Collection gutters along walls are usually 2"-6" wide and center floor gutters are 8"-12" wide. Limit floors to one or two sloping surfaces to reduce ponding, Fig 5-10.

Milk room drains must handle milking equipment and bulk tank washwater. For easy access and good sanitation, do not locate floor drains under the bulk tank or its outlet valve. Size drains and pipelines to handle the water from cleaning and sanitizing operations. Usually, 4" diameter pipelines can be used for branch lines and 6" diameter pipelines for main lines.

The amount of solids that can be handled by the waste disposal system affects the design of the cow platform drainage system. A flat floor without gutters is easier to clean and reduces the amount of solids washed into the system. For easier cleaning and washdown, slope the floor along the length of the parlor toward the entrance to a cross gutter, Fig 5-11a. If gutters are used along the length of the parlor, slope the floor toward the entrance with a cross slope from the pit to the gutter, Fig 5-11b. Slope the gutter bottom to drain toward one end.

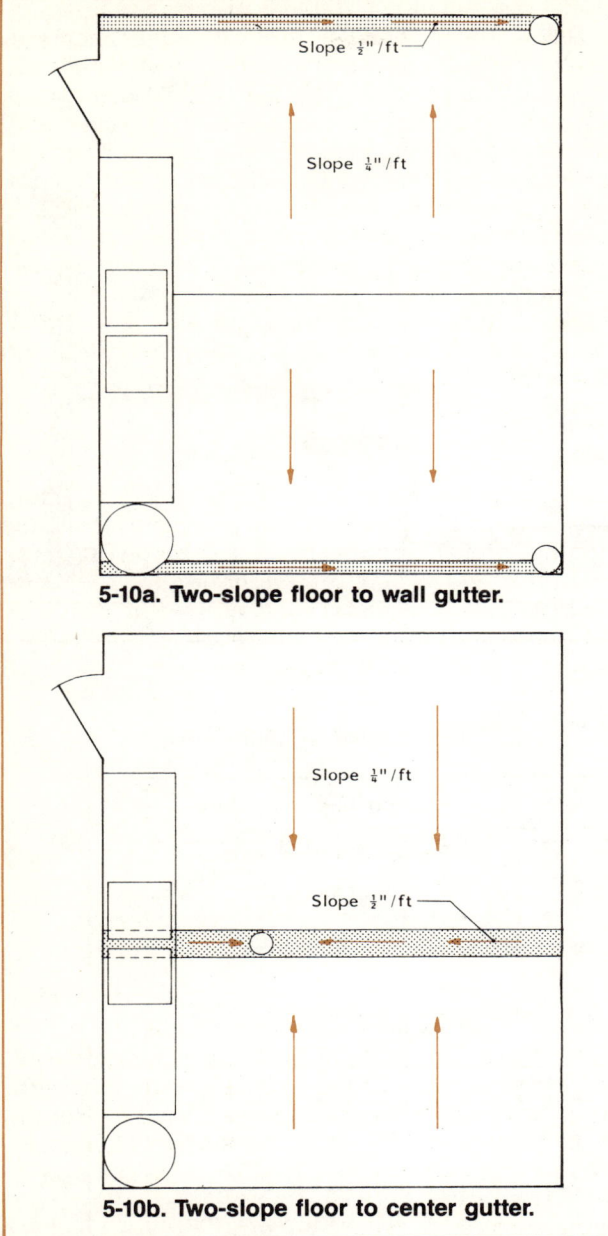

5-10a. Two-slope floor to wall gutter.

5-10b. Two-slope floor to center gutter.

Fig 5-10. Milk room floor slopes.
Slope floors a minimum of ¼"/ft.

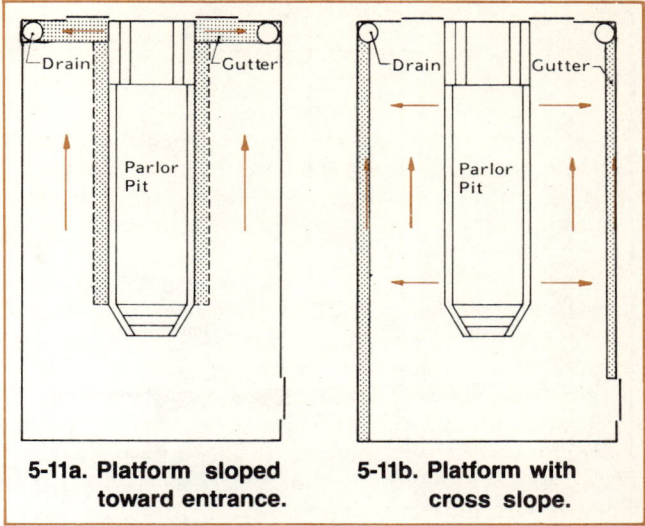

5-11a. Platform sloped toward entrance.

5-11b. Platform with cross slope.

Fig 5-11. Milking platform drain system.

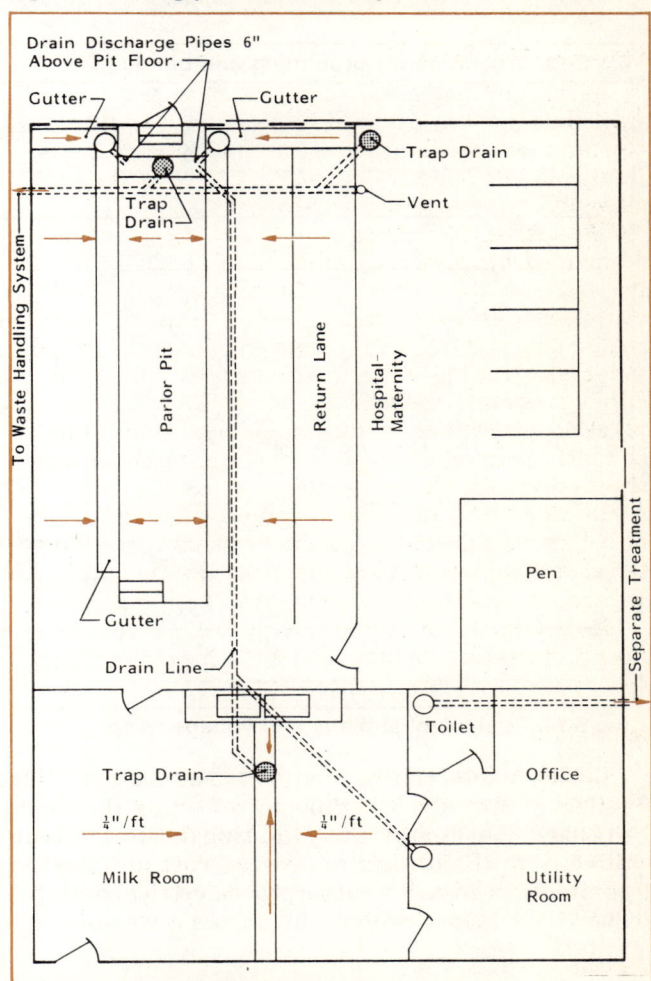

Fig 5-12. Milking center drainage system.

Plan ahead to minimize the number of deep-water-seal drains needed. Install one drain in the milk room and another in the gutter at the end of the operator pit. The parlor, cow platform, and milk room drains can discharge through the pit wall about 6" above the pit gutter. Bringing these drains through the pit wall allows for observing the drains for problems, Fig 5-12.

Lighting

Provide two continuous rows of 40 W cool white fluorescent tubes the length of the milking parlor. Install each row over the operator's pit 10" from the platform edge. Avoid equipment that interferes with lighting.

Provide one waterproof fluorescent fixture with two 4'-40 W cool white fluorescent tubes for each 100 ft² of floor area and an additional fixture over each vat. Locate fixtures so broken glass does not fall into bulk tank openings. Adequate lighting to the bulk tank interior makes individual spotlights unnecessary.

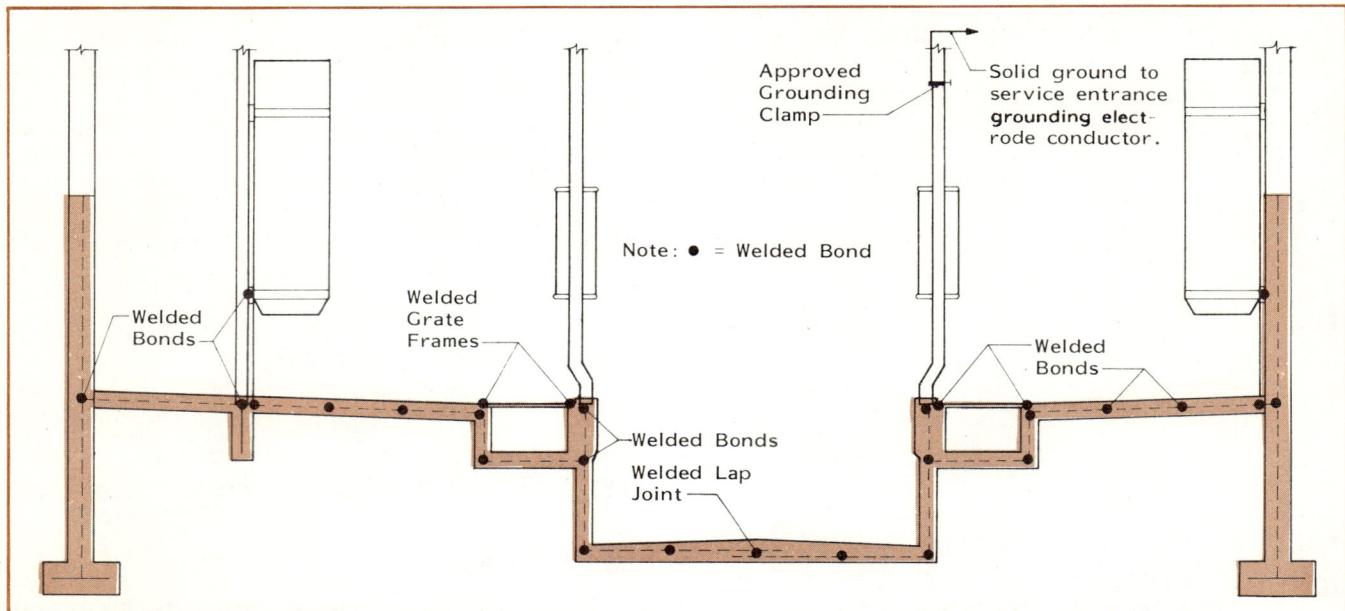

Fig 5-13. Milking parlor grounding and bonding.

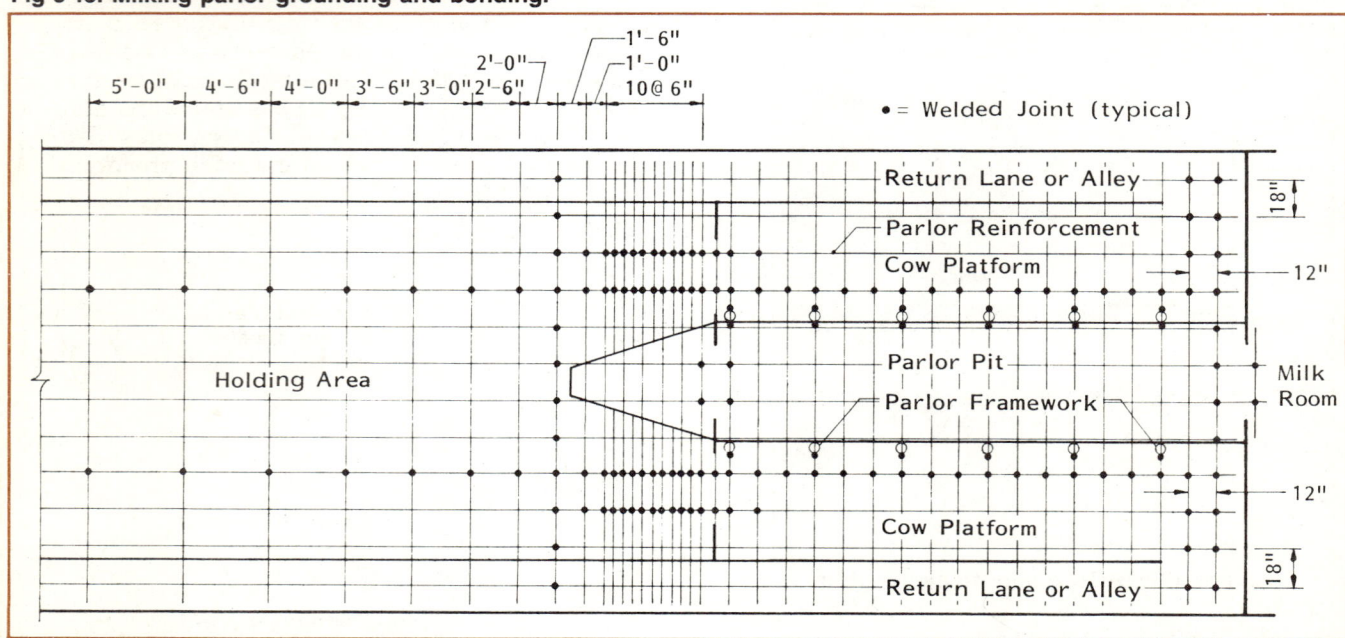

Fig 5-14. Equipotential plane and voltage ramp.

With white ceiling and wall surfaces, no light reflectors are needed. Provide reflectors if surfaces are dark. Use weatherproof lighting fixtures fastened directly to the ceiling to prevent rust and dust accumulation. Install weatherproof electrical outlets 5′ above the floor to protect them from cows and washwater.

Stray Voltage

Stray voltage is a small voltage difference that exists between two surfaces (stanchion, waterer, floor, etc.) that a cow can reach at the same time. When a cow touches both surfaces, current flows through her body. Voltage differences as low as 0.5 V can adversely affect cow movement and milk production and increase incidences of mastitis. All producers must minimize the potential of stray voltages.

Stray voltages are often caused by faulty wiring or problems in the grounded neutral network. The solution can be simple if the problem sources are clearly diagnosed and the alternatives evaluated. Good diagnosis requires a thorough understanding of electricity and farmstead wiring fundamentals. If you feel you have a problem, contact your electrician, power supplier, and state extension agricultural engineer for assistance.

In new dairy installations, interconnect all steel within the milking center with welded connections to form an equipotential plane. An equipotential plane maintains equal voltages at all points in the milking center to eliminate current flow. Bolted, wired, or clamped connections are not satisfactory. Before placing concrete in the milking parlor, weld all parlor equipment supports to the reinforcing steel. Use No.

3 rebars for the reinforcing steel. Because of installation difficulties, welded wire mesh is not recommended.

With approved clamp-type connectors and No. 6 copper wire, electrically ground the milking center to the ground at the distribution panel. Fig 5-13 shows typical weld connection locations and grounding for a milking parlor.

Access to the equipotential plane should be through a voltage ramp installed in the holding area. Voltage ramps provide a gradual voltage rise to the equipotential plane voltage. Variable-spaced steel No. 3 rebars make up the voltage ramp gradient. See Fig 5-14 for recommended spacing. Interconnect all bars across the holding area to at least two longitudinal bars with welded joints.

Extend at least two bars from the holding area through each parlor entrance. Interconnect individual mesh sections with at least two welded connections to form a continuous metal grid. Fig 5-14 shows a recommended layout for a holding area voltage ramp and the interconnection of the grounding network.

Milking Center Environment

Each room in the milking center has its own environmental requirements. Consider milking center activities and heating, ventilating, and cooling principles when designing an environmental control system.

Holding Area

The ventilating system for the holding area must remove animal moisture and heat. With natural ventilation, provide an open ridge and open eaves for winter moisture removal. Add sidewall openings for summer ventilation. With a mechanical ventilating system, base fan capacity on the animal holding capacity and recommended ventilating rates. See the building environment chapter. Install proper air inlets and controls.

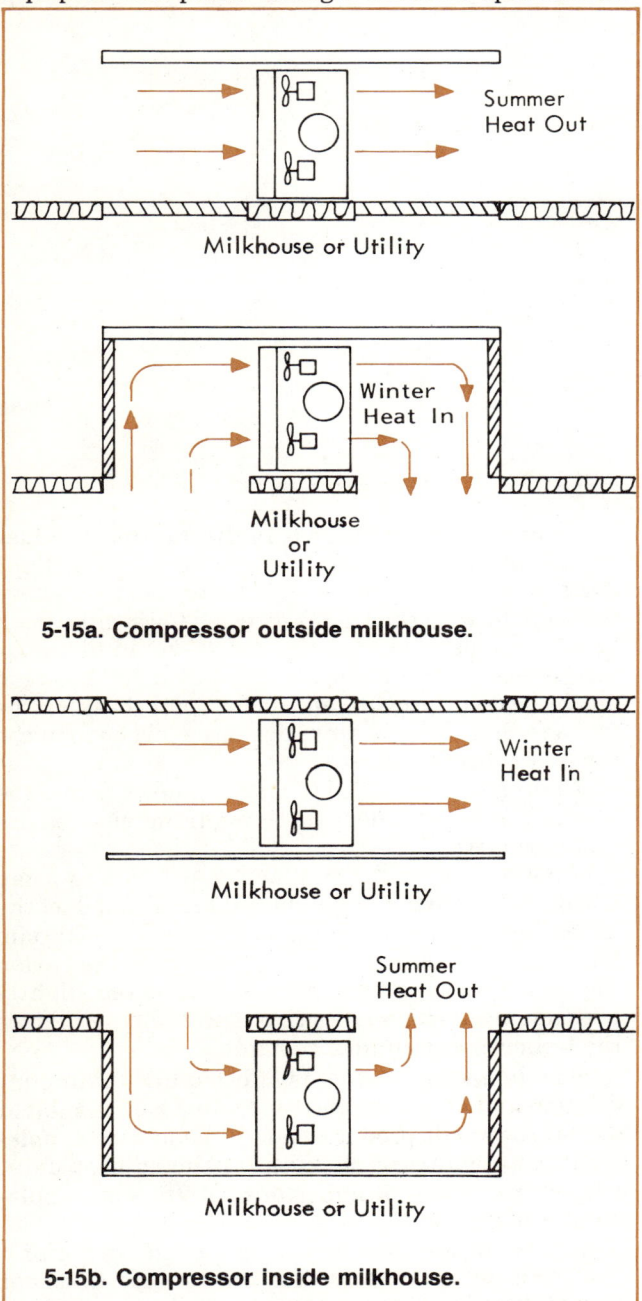

5-15a. Compressor outside milkhouse.

5-15b. Compressor inside milkhouse.

Fig 5-15. Compressor sheds.
The doors retain compressor heat in cold weather and open to remove it in warm weather. Make the cross-sectional area of all air passages 1½ times the size of the condenser area.

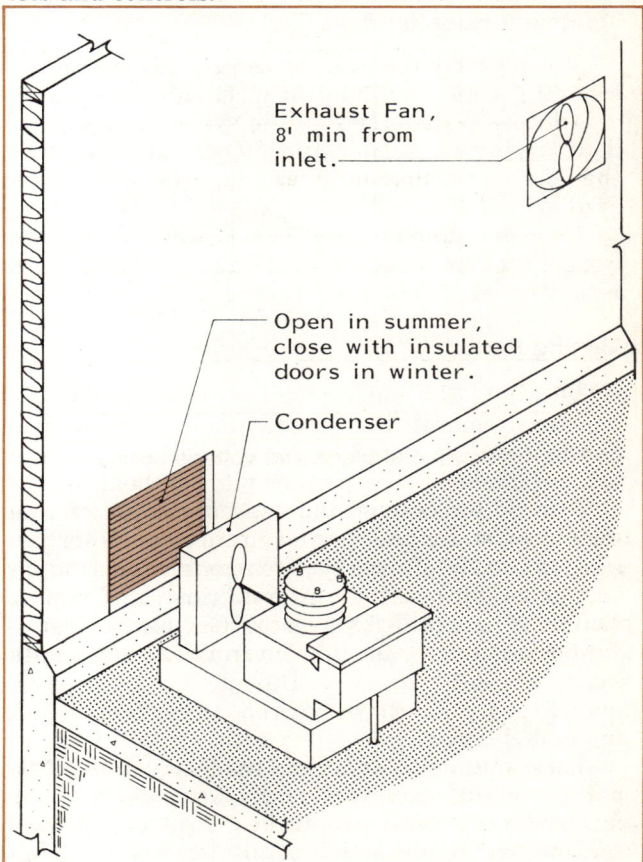

Fig 5-16. Compressor ventilation.

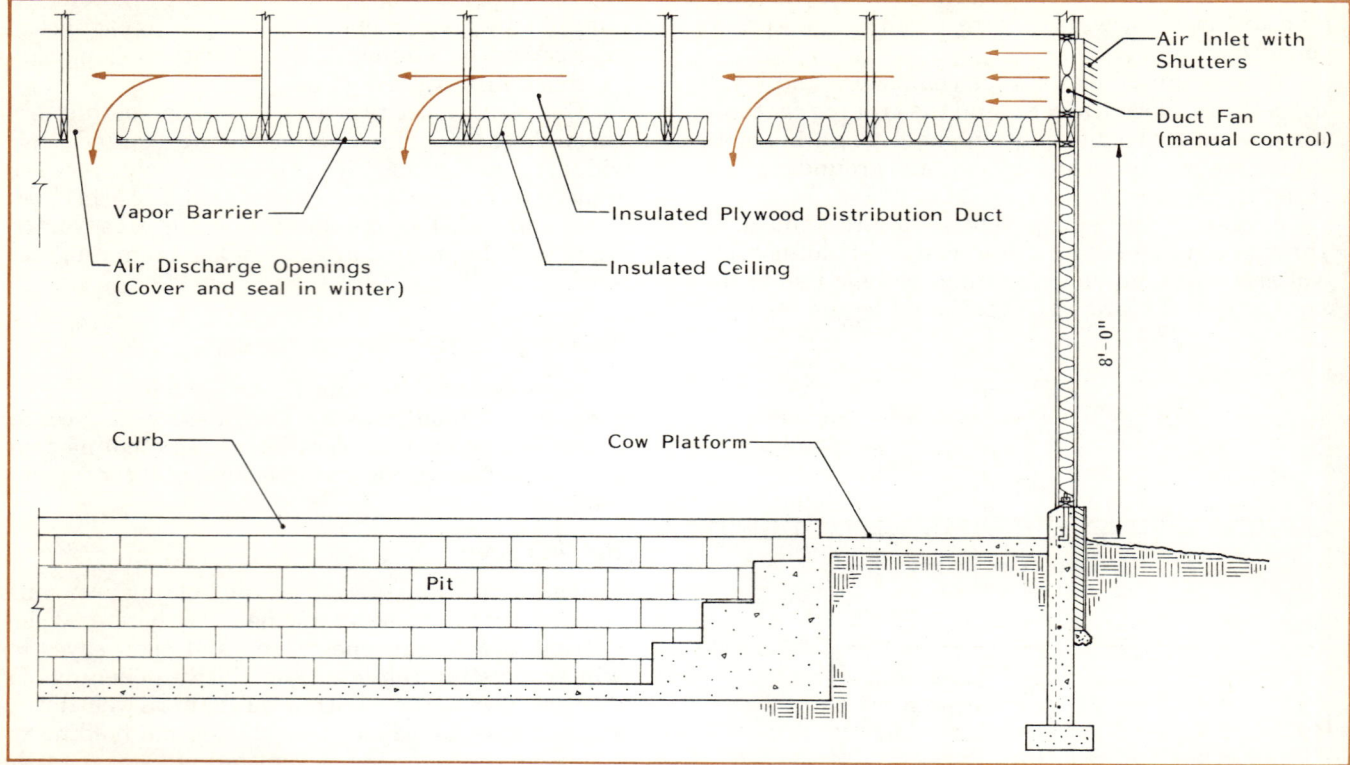

Fig 5-17. Summer air duct
Provide at least 2 openings to the milking parlor.

Treatment-Hospital Area

When part of the milking center, this is usually a separate room with its own mechanical or natural ventilating system. Design the system to accommodate varying animal densities. If pens are adjacent to the holding area, use the holding area's ventilating system.

Consider supplemental heat if water freezing or worker comfort are concerns. Use an individual space heater or central heating system.

Milking Parlor

Ventilate the parlor for operator comfort and moisture removal. Use exhaust fans to remove heat and moisture given off by cows, circulating fans for air movement in the operator pit, and heaters.

Provide an exhaust fan capacity of 100 cfm per stall at ⅛" static pressure for minimum winter ventilation and 400 cfm per stall for summer ventilation.

Control fans with an interval timer and manual switch in parallel for greater flexibility. During winter, use a timer to provide ventilation after milking to dry out the parlor. During summer, use the manual switch for continuous operation during milking and cleanup.

Locate the exhaust fans on the leeward side of the parlor and exhaust directly outdoors, away from any milkroom or office air intakes. Provide ½" mesh screened air inlets with a total net area of 1 ft^2 for each 600 cfm of fan capacity. Locate the fan(s) and air intakes to ensure air movement through the entire parlor area.

Increasing air movement in the pit area provides additional summer operator comfort. Underfloor ducts to the milker's pit should enter the pit 8" above the pit floor with the hot air directed toward the floor. Insulated plastic or clay bell tile 8"-12" in diameter can be used for the ducts. Locate cold air returns high on the wall to provide an airflow path back to the furnace. Size the cold air return area larger than the warm air discharge area.

Ceiling fans mounted to move air down across the operator or fans to blow air through the pit area are an alternative.

Another option for summer ventilation is a fan and duct. Locate the fan in the gable end and duct the air to the parlor. Provide 1 ft^2 of duct for each 600 cfm fan capacity. Use two or more openings to the parlor, Fig 5-17. Provide a total net discharge area slightly less than the duct cross-sectional area. For details see the building environment chapter.

Supplemental heating is affected by building insulation and operator preference. In a well insulated parlor, cows will produce enough heat. Use supplemental heat during milking on the coldest days. Proper insulation is important to efficient supplemental heat use.

The heating system for the parlor can use radiant, floor, unit space, or central heating systems. Consider the total milking center heating needs when selecting the parlor heating system.

If possible, use heat removed from the milk to heat the milking center. A central forced air or hot water heating system can allow for this. See utility room section.

Use radiant heaters for heating the operator without heating the air. Hang heaters from the ceiling and direct toward the cow's udder and the milker's hands. Additional heaters can be installed over other work areas.

Milk Room

The environmental control requirements are moisture removal, minimized heat buildup, and keeping equipment from freezing. Because this is the room where milk is stored and equipment is cleaned, take care to keep odors and dust out.

A pressurized ventilating system with filters on the inlet to the fan is preferred. Locate the fan away from sources of excess dust, odors, and moisture. The pressurization minimizes the opportunity for odor to enter from the parlor. A 600 to 800 cfm fan is sufficient. A larger fan is needed if compressors are located in the tank room. Provide heat in the milk room with a unit heater or central heating system to prevent freezing.

Office

Use the central heating system or a space heater to heat the office and toilet. Heater size depends on building insulation levels and winter design temperatures. Install a small exhaust fan in the toilet room. Consider installing additional summer ventilation or air conditioning in the office.

Storage Room

Maintain storage room temperatures between 40-80 F. Low temperatures can damage medical supplies while high temperatures can accelerate rubber component deterioration. Maintain safe storage temperatures with a supplemental heater or duct from a central heating system. Do not use heat from the utility room.

If an interior space is used and all surrounding walls are insulated, an additional heat source may not be needed. Heat from refrigerators used to store rubber products and pharmaceuticals will keep the space warm. Provide a source of fresh air and a thermostatically controlled exhaust fan to control temperature rise. Do not bring air from dusty driveways or animal housing into the storage room.

Utility Room

The utility room must not freeze during the winter. Usually, the equipment produces enough heat to meet this need. Provide adequate ventilation for efficient equipment operation.

Move the heat from air-cooled compressors to other rooms with a fan and duct.

Use an induced draft or powered vent furnace to force exhaust gases out powered dampers. Select a furnace with easily changeable filters.

A central hot water system can heat each room with convectors, Fig 5-19. The heated water from the milk cooling unit can be used as a hot water source. Allow for individual temperature control in each room. Use a hot water heater or a boiler for supplemental heat when needed. Local regulations may not allow this system if the same heat exchanger is used in preheating the washwater.

Regardless of the methods of using the waste heat in winter, provide a thermostatically controlled exhaust fan that turns on at a preselected temperature. Select a fan to provide one air exchange per minute. Provide 1 ft^2 of air inlet area for each 600 cfm of fan capacity. Cover the air inlets with gravity or motorized louvers that open when the fan turns on.

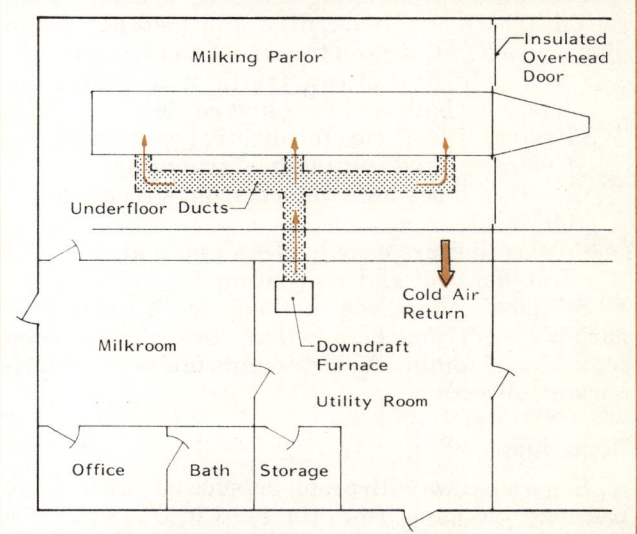

Fig 5-18. Forced air central heating system.

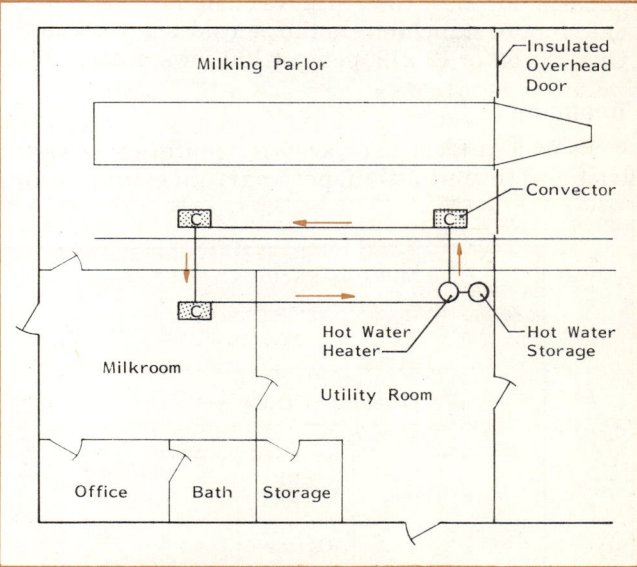

Fig 5-19. Hot water central heating system.

6. SPECIAL HANDLING AND TREATMENT

Some means of restraining cattle for medical examination, treatment, hoof trimming, artificial insemination, and other procedures is needed. Provide a separate area for animals with long-term illnesses.

Do not examine or treat cows in the milking parlor. It is not designed for these purposes. Cows may resist entering the milking parlor if afraid of being treated.

In stall barns, cows are often treated in the stalls, but a separate handling facility for easier restraint, examination, treatment, and surgery is recommended.

Four different handling areas are needed—separation, treatment, maternity, and loading. When planning a handling and treatment facility, provide:
- A smooth flow of cow traffic with maximum safety for both handlers and cattle.
- Restraint facilities for medical examinations, treatment, and routine hoof trimming.
- Parking space for a veterinarian's truck, possibly inside.
- Areas that are easy to clean and maintain.
- Suitable heat and ventilation.

Systematic handling methods speed the veterinarian's work and help collect the animals, keep records, and administer treatments under the veterinarian's direction.

Separation

Separate cows with problems such as sore feet and bad legs and joints from the herd in free stall systems. Pen them near the milking parlor, a feed bunk, waterer, and manure alley if possible. A dirt floor may be easier for these cows to move on in the pen. Provide one or more stanchions in the pen sides as restraints. One 10'x14' or 12'x12' pen per 100 cows is desirable.

Treatment

Provide a treatment area for confining cows for artificial insemination, post-partum examination, pregnancy diagnosis, sick cow examination, and surgery. Allow a 3'x5' area in front of and behind a cow to perform a complete examination. Locate the treatment area in the milking center where hot and cold water, heat, and refrigerated storage are available.

An area alongside the holding area provides a convenient place to sort animals and to perform medical examinations and treatments, Fig 6-1. Sort cows exiting the parlor into a catch lane with a sorting gate controlled from the pit. They can be held there or can be moved into one of the pens equipped with self-locking stanchions. With gates to block the parlor entrance, a group of animals can be moved through the holding area into the treatment area for sorting, examination, or treatment.

For large herds, a separate veterinary cattle handling facility is desirable, Fig 6-2. A separate facility allows for more efficient cattle handling, examination, and treatment. Angled restraining stalls permit the veterinarian unrestricted access to the rear and sides of a cow. After cows are released, they can be returned to the barn or diverted to a hoof trimming table.

In stanchion or tie stall barns, a single stall can be used for treatment, Fig 6-3. Locate the stall in the tie stall line where neighboring cows can be moved or next to a cross alley, for better access to the cow during examination.

Restrain the cow's head with a swing stanchion that can lock into position with a removable bar. Use a sturdy ring in the wall in front of the stall for further restraint. Provide 3 rings in the ceiling over the centerline of the stall for lifting points. Locate one lifting point above the cow's shoulders, another above the tailhead, and a third 4'-6' behind the cow. Lifting point rings must be strong enough to support the cow's weight.

Gang Stanchions

In group housing gang stanchions along a feed manger can restrain animals for heat detection,

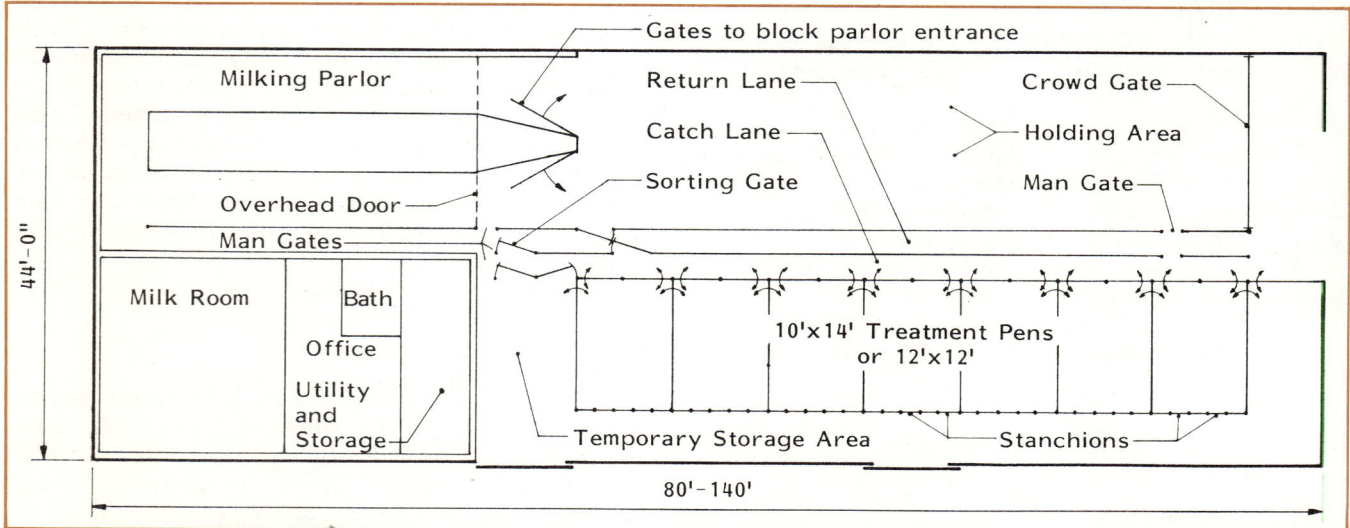

Fig 6-1. Separate treatment area.

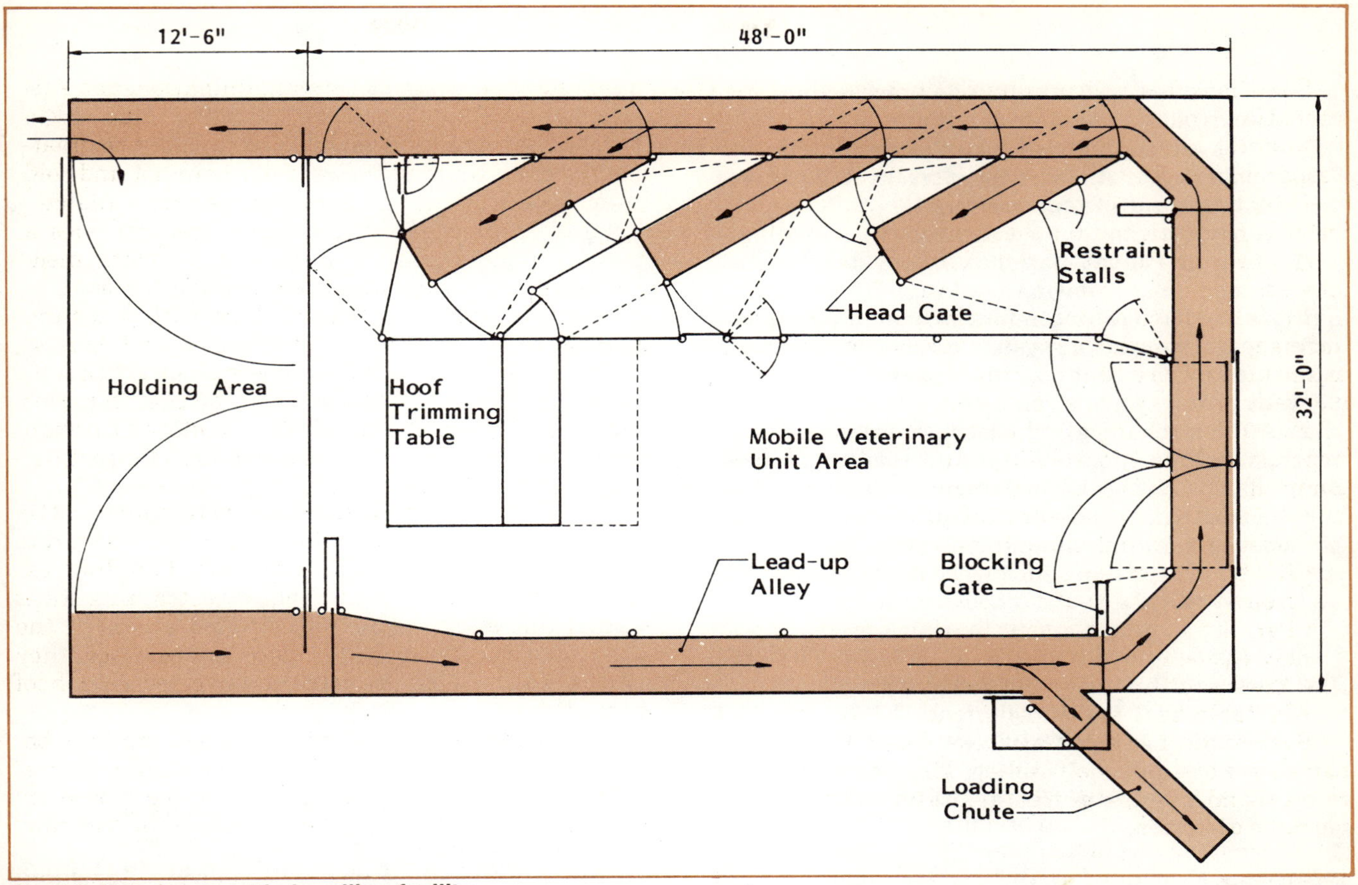

Fig 6-2. Veterinary cattle handling facility.

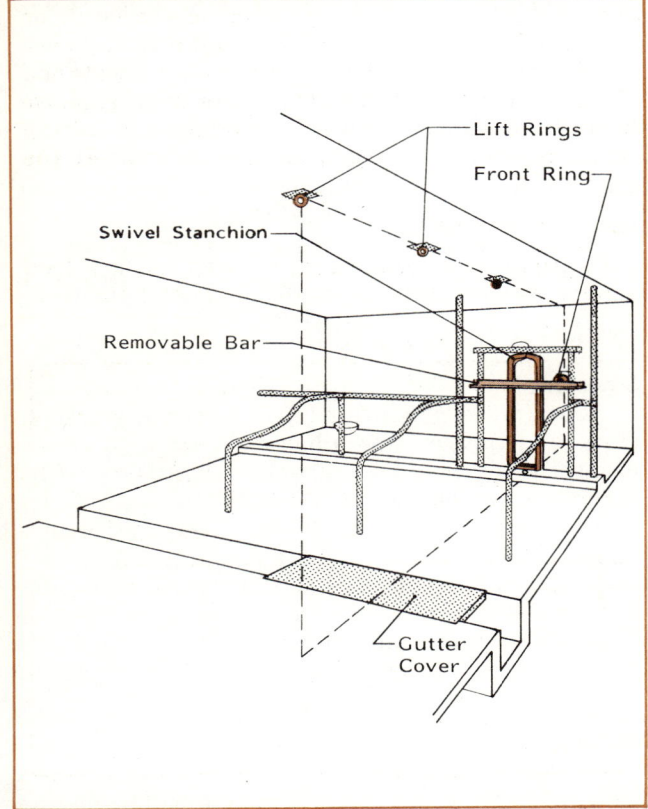

Fig 6-3. Treatment stall.

breeding, pregnancy-checking, etc. Gang stanchions permit selective feeding. A lever mechanism opens or closes all stanchions simultaneously. Desirable features are individual or group cow release, self-locking stanchion as a cow lowers her head after entering, and a quick release for downed cows. For mature cows place stanchions at least 2' o.c. Provide at least the same number of stanchions as cows in the group.

Maternity Area

Make maternity pens wide enough for a cow to turn around easily and long enough to use a calf puller when needed. Recommended sizes are 12'x12' or 10'x14'. Provide one maternity pen for every 25 cows. Place a stanchion diagonally across one corner of the pen. A concrete curb between each stall aids sanitation. Provide dirt floor maternity pens for cows insecure on concrete. If maternity pens are in the milking barn, do not use dirt floor pens. Use deep bedding on concrete floors to prevent cows from slipping.

Locate pens for easy access with a tractor or loader.

Loading Chutes

Provide a chute with solid sides for receiving and shipping cows. Provide a holding pen to hold cows before loading. Provide steps or cleats on the chute floor so the cows do not slip. See the equipment chapter for examples of loading chute construction.

7. BUILDING ENVIRONMENT

Ventilation is an air exchange process that:
- Brings outside air into a building.
- Moves air through all areas of the building to supply oxygen and pick up odors, heat, moisture, dust, and pathogenic organisms.
- Removes contaminated air from the building.

The purpose of ventilation is to provide a healthy environment. Base ventilation design on animal requirements. If operator comfort is considered, adjustments in the ventilating system can be made, but the resulting environment may not be best for livestock.

Ventilating systems can be either natural or mechanical. A mechanical ventilating system uses fans, controls, and air inlets or outlets to provide a positive air exchange. Natural ventilation utilizes wind pressure and differences between inside and outside temperature to move air through the building.

Housing Types

The environment within a building can be classified based on the winter temperature.

A **cold barn** has indoor winter temperatures about the same as outside temperatures. It is usually insulated and has unregulated natural ventilation. Dairy animals do satisfactorily in properly ventilated cold barns if kept dry and properly fed. Proper equipment selection and management minimize the inconveniences associated with freezing weather. Cold barns are usually naturally ventilated but can be mechanically ventilated.

A **modified environment barn** has indoor winter temperatures higher than outdoors and usually above 32 F. Modify winter temperature by adding insulation and adjusting the openings in a naturally ventilated building. These barns are generally naturally ventilated.

A modified barn has fewer problems with frozen manure. Insulation increases the cost of modified housing but helps keep the barn cooler in summer and warmer in winter.

Warm barns are kept at 40 F or above by good construction, careful control of ventilating system, and adding supplemental heat as needed. Warm barns are usually stanchion barns, calf barns, and milking centers that are mechanically ventilated, but can be freestall barns.

Natural Ventilating Systems

Principles of Natural Ventilation

Wind pressure and the difference between inside and outside temperature moves air through the building. Natural ventilation of gable buildings works best with a continuous ridge opening, large sidewall openings, no ceiling, and small continuous eave openings, Fig 7-1.

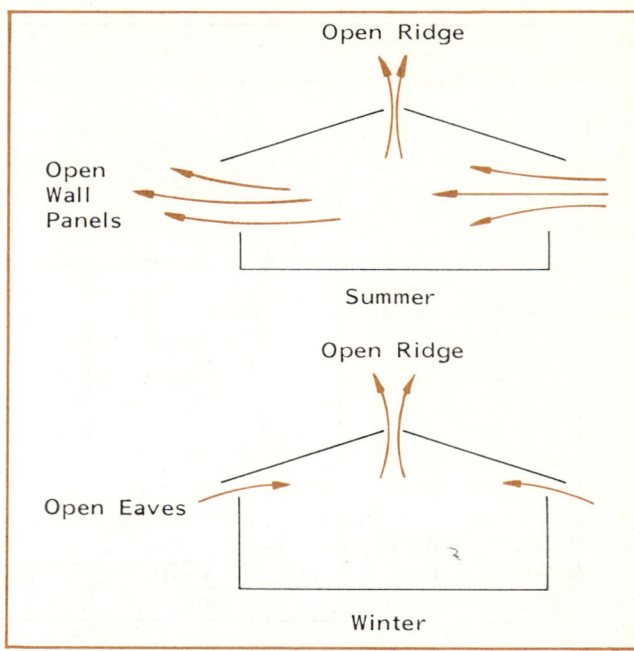

Fig 7-1. Naturally ventilated barn features.

Winter ventilation

Wind across the open ridge along with inside and outside temperature differences draw warm, moist air out through the ridge and fresh air in through continuous sidewall or eave openings. Locate winter air inlets high on the sidewall or in the eaves. Downwind air inlets will occasionally act as air outlets. Some ventilation occurs even on calm days because warm air rises causing a chimney effect.

Summer ventilation

Large openings (typically ½ or more of the sidewalls) are important to good summer ventilation. Sidewall openings and end doors allow a cross flow of air. Locate openings to ensure good airflow through the animal zone (about 4' above floor level).

Design

Site selection

Trees, tower silos, grain bins, and other structures disturb airflow around adjacent buildings. Locate naturally ventilated buildings on high ground for better wind exposure and at least 50' in any direction from other structures and trees. Trees or structures affect both summer and winter ventilation for 5 to 10 times their height downwind. If naturally ventilated buildings are in a building complex, place them on the west or south side of mechanically ventilated buildings and perpendicular to prevailing summer winds.

Building orientation

Building orientation affects the performance of a natural ventilating system. Greater wind pressure differences occur when air strikes the side of a building rather than the endwall. Because air enters at the eave or sidewall openings and discharges at the ridge, orient buildings so prevailing winds are perpendicular to the ridge openings. This is especially important in warm weather to maximize cross airflow.

In the Midwest, build open-front buildings with the long axis east-west for the best summer cooling, winter sun penetration, and winter wind control.

Ridge openings

Provide a ridge opening of 2" (measured horizontally) for each 10' of building width, e.g. a 14" wide opening for a 70' wide building.

To protect building framing, cover each rafter or truss with a strip of flashing material. Extend the flashing ½"-1" on each side of the truss member or rafter, Fig 7-2.

With a properly sized ridge opening, precipitation entering the building is not a serious problem. Air exiting through the ridge prevents most rain and snow from entering. Raised ridge caps are not recommended because they disturb airflow, can trap snow, are expensive, and require maintenance. An upstand, Fig 7-3, can be added to help keep out snow and rain.

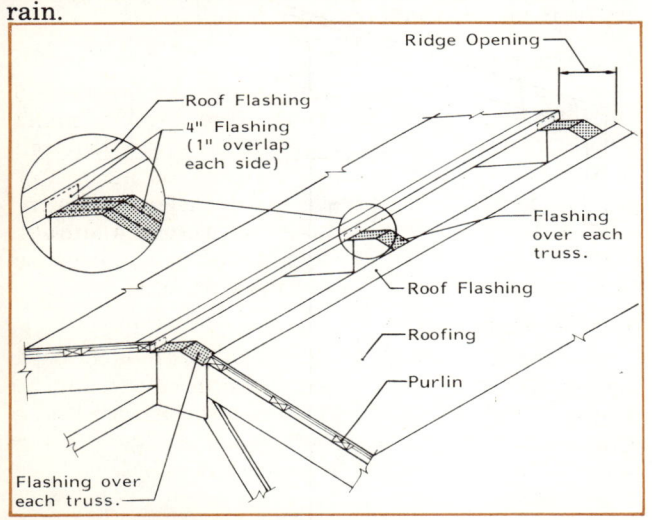

Fig 7-2. Flashing for ridge openings.

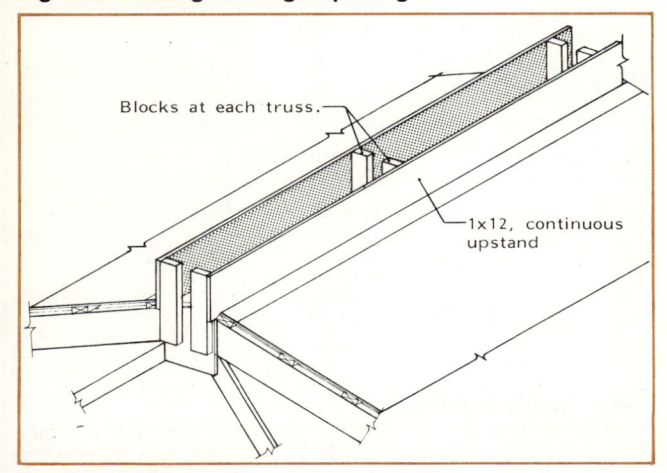

Fig 7-3. Upstand.
12" upstand or baffles help keep out wind and snow and also increase chimney effect when the wind is blowing perpendicular to the ridge.

When possible, locate scraper alley or other less critical areas under the ridge. If the amount of precipitation coming through the ridge is objectionable, it is better to protect areas below the ridge than to build a cap. Cover critical components, such as a feeder motor or feeder belt, or place an internal gutter 2'-3' below the open ridge to collect rain water and channel it out of the building, Fig 7-4.

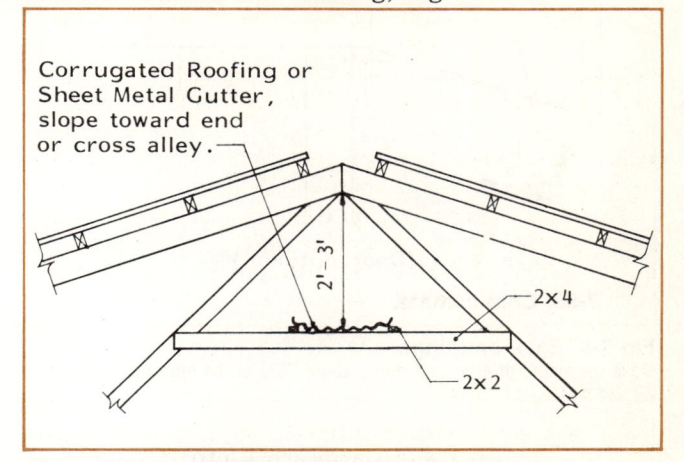

Fig 7-4. Interior trough with an open ridge.

Eave openings

Construct continuous eave openings along both sides of the building. Size **each** opening to have at least ½ as much open area as the ridge opening. Eave openings can be provided at the open spaces between trusses or on the underside of the roof overhang, Fig 7-5.

Provide eave baffles so airflow can be reduced, but never completely close them. To reduce drafts, protect eave openings from direct wind gusts with fascia boards and vent doors. Use vent doors only during periods of severe winter weather. If vent doors are used, continual adjustment is needed requiring more labor.

Sidewall openings

Larger sidewall openings are needed for cross ventilation during warm weather. Provide openings in the closed side of open front buildings and both long walls in totally enclosed buildings.

Several types of summer vent doors can be used—pivot doors, top- or bottom-hinged doors, sliding doors, and plastic or nylon curtains, Figs 7-6 and 7-7. Provide at least a 4' high continuous opening the length of the building. For good airflow at the cow level, put the lower edge of the opening about 4' above the floor. Build vent doors to open fully so none of the opening is constricted or increase door size to provide the recommended opening.

Make sidewalls 10' high to take advantage of winds for ventilation. Higher sidewalls are sometimes needed for maneuvering equipment, but they are no better for the ventilating system.

End walls

Provide ventilating openings in the end walls of wide buildings. Alley doors may be adequate. Provide removable ventilating panels or wall sections in buildings with no large end doors.

7.3

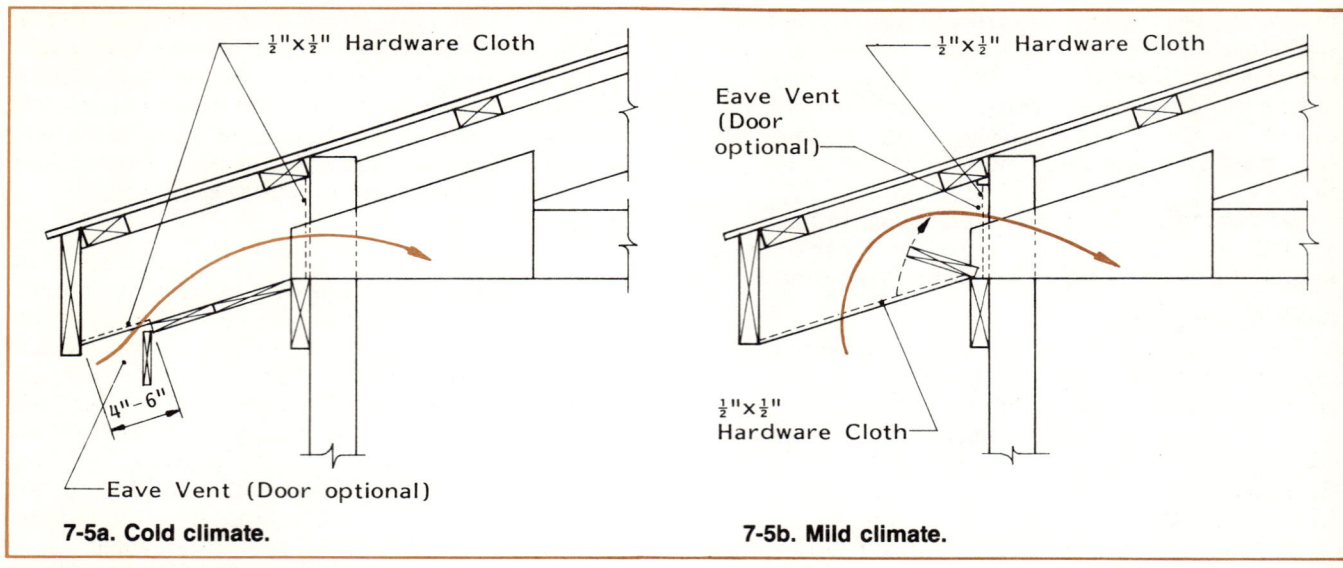

7-5a. Cold climate. **7-5b. Mild climate.**

Fig 7-5. Eave openings.
Make the length of eave vent doors about 75% of the opening length.

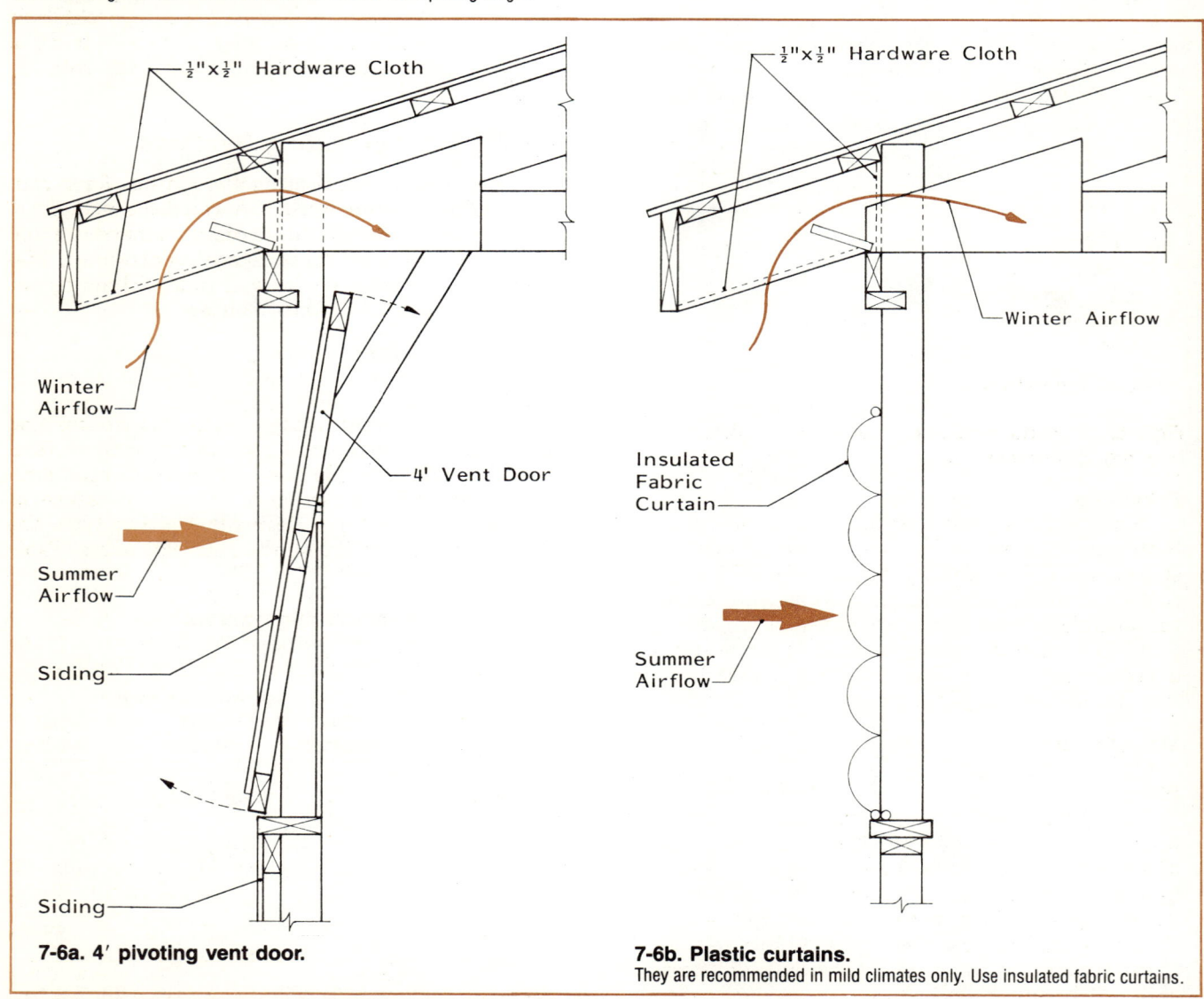

7-6a. 4' pivoting vent door. **7-6b. Plastic curtains.**
They are recommended in mild climates only. Use insulated fabric curtains.

Fig 7-6. Air inlets, cold and modified environment housing.

Poor winter ventilation and high humidity cause animal health problems. Fog, condensation, or frost can form when the building is not ventilated properly. Attempts to warm the building by closing eave inlets, wall openings, or ridge openings with baffles increase the problem. Always provide at least a ½" continuous eave opening per 10' of building width.

Summer

Hot, muggy summer weather is one of the most critical times for animal health, comfort, and productivity. Often there is little air movement and sunshine on the roof radiates more heat to the animals.

Provide recommended sidewall openings for good cross ventilation to reduce animal heat stress. Adequate ridge and eave openings are important for proper air movement through the building. Low, wide buildings and buildings with flat ceilings or low pitched roofs are difficult to naturally ventilate because adequate openings cannot be provided.

Modify existing single-story buildings with poor ventilation by removing an existing ceiling, opening a ridge, and cutting large sidewall openings for better natural air movement. If summer heat stress is still a problem, use circulation fans to move air over the animals.

Mechanical Ventilating Systems

A mechanical ventilating system has fans, controls, and air inlets or outlets. A well designed system provides greater control over room temperature and air movement than does natural ventilation. Mechanical ventilation works well in stanchion barns, calf barns, and the milking center.

Design Principles

Ventilating rates

An effective ventilating system distributes and mixes the air within the building to control temperature, moisture, odor, and pathogenic organisms. The system can be an exhaust (negative pressure) or positive pressure type. Regardless of the type, the recommended ventilating rates are the same, Table 7-1.

Table 7-1. Dairy barn ventilating rates.
Size the system based on total building capacity. Table values are additive—e.g. for calves, mild weather requires 15 + 35 = 50 cfm/calf.

	Cold weather*	Mild weather	Hot weather
	--------- cfm/animal ---------		
Calves 0-2 mo	15	+35 = 50	+50 = 100
Heifers			
2-12 mo	20	+40 = 60	+70 = 130
12-24 mo	30	+50 = 80	+100 = 180
Cow 1,400 lb	50	+120 = 170	+300 = 470

*An alternative cold weather rate is 1/15 the room or building volume; ft³/15. An alternate hot weather rate is the building volume divided by 1.5.

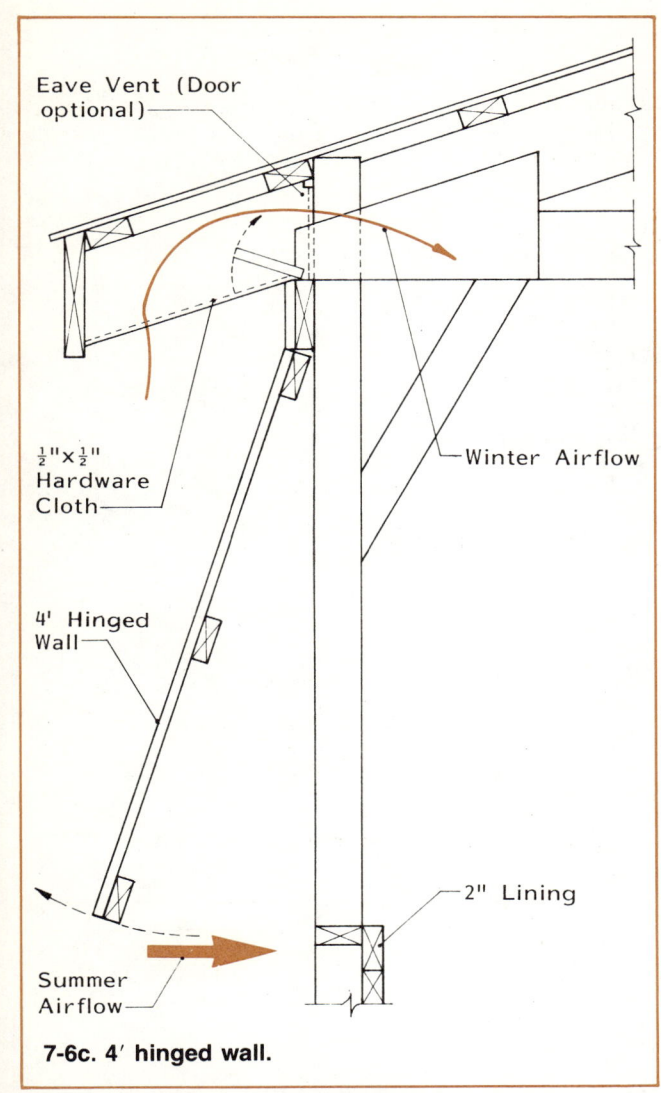

7-6c. 4' hinged wall.

Fig 7-6. Air inlets, cold and modified environment housing, continued.

Roof slope

Since warm air rises, steeper sloped roofs provide better upward warm airflow. However, with roof slopes over 6/12, incoming air rises rapidly along the roof and does not drop into the animal zone. Condensation and high interior summer temperatures are a problem with roof slopes less than 4/12 because of reduced air movement. Therefore, for livestock housing, use a 4/12 to 6/12 roof slope.

Management

Winter

Use water bowls and pipes that are heated or frost protected. During cold weather, remove manure often. Scrape the barn immediately after cows are moved to the holding area. Scrape the barn twice a day if necessary. In extremely cold weather, scrape in the afternoon when conditions are slightly warmer. A heavy tractor scraper or bucket with down pressure will help break up frozen manure.

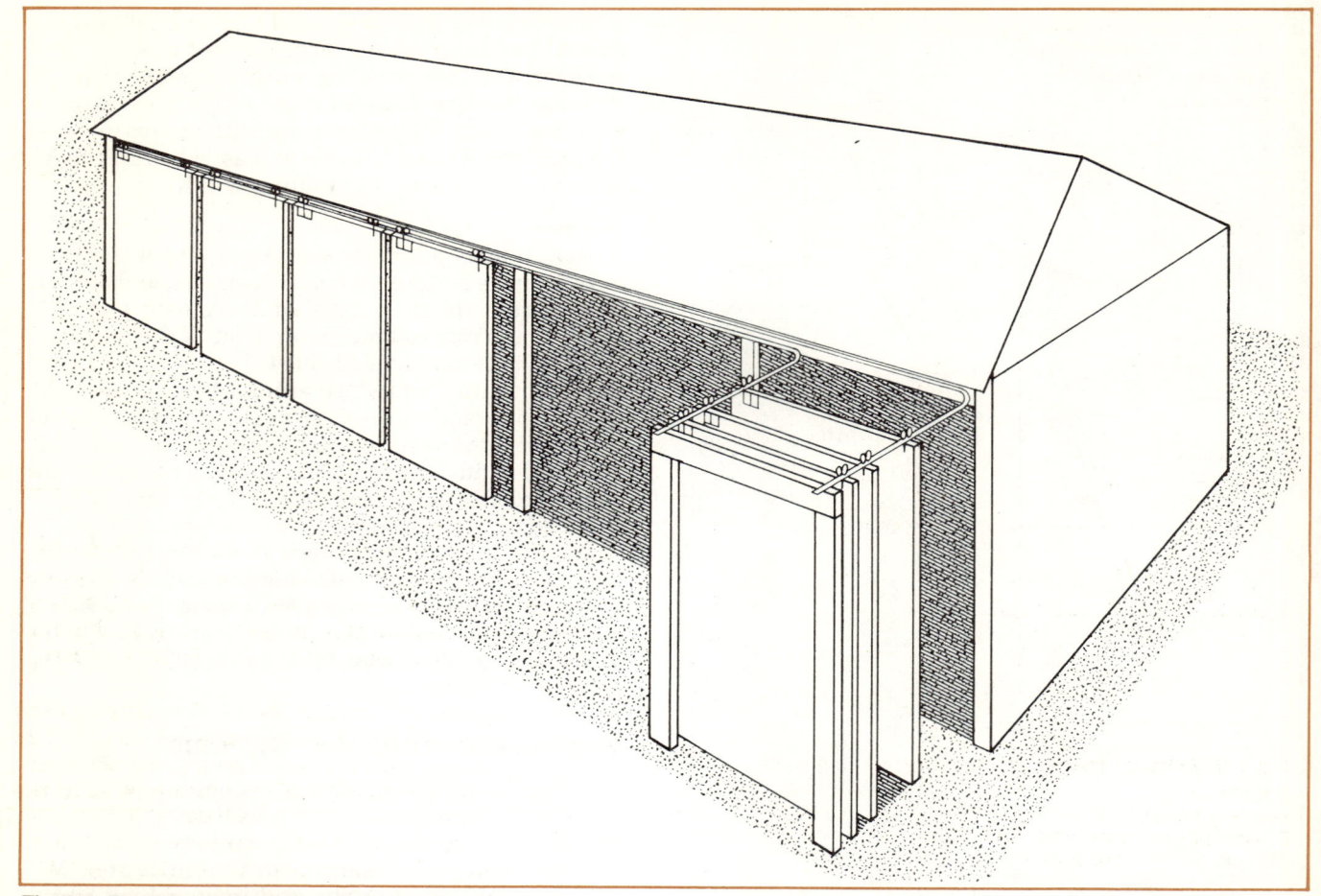

Fig 7-7. Double track sliding door system.
Large sidewall openings help increase summer ventilation.

Temperature control

To maintain a constant room temperature, the heat **produced** by animals, electric equipment, and heaters has to equal heat **lost** through building surfaces and ventilation. Supplemental heat is required in calf barns and milking centers. The heat production of a 1,300 lb cow can be from 3,800 Btu/hr at 30 F to 1,300 Btu/hr at 80 F.

Moisture control

A 1,300 lb cow can release between 3 gal/day at 30 F to 7 gal/day at 80 F of water vapor. This moisture must be removed by the ventilating system.

During cold weather, ventilation brings cold, relatively dry air into the building. The air is warmed by energy from animals, electrical equipment, and supplemental heat. As the air temperature rises and the air expands, its relative humidity decreases and it has a higher moisture holding capacity. The moisture holding capacity of air nearly doubles for every 20 F rise in temperature. The ventilating air picks up moisture, which increases the relative humidity, and the system exhausts the ventilating air and moisture from the building, Fig 7-8.

Design ventilation to maintain room air at 50%-75% relative humidity. Higher humidities increase condensation; lower humidities increase dust levels. Also, 50%-75% relative humidity reduces airborne bacteria found in livestock buildings.

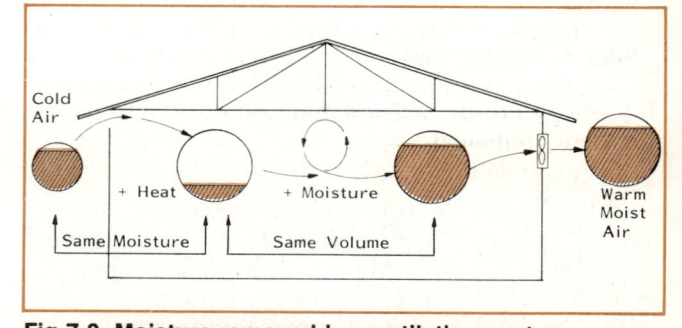

Fig 7-8. Moisture removed by ventilating systems.

Ventilating System Types

Exhaust ventilation

Exhaust fans pull air out of the building, creating a vacuum or negative pressure inside the building. The pressure difference creates airflow through designed inlets and other openings, Fig 7-9. If the inlets are sized and located properly, fresh air is distributed uniformly throughout the room. Size exhaust fans based on Table 7-1. Exhaust is by far the most common mechanical ventilating system in dairy housing.

Positive pressure ventilation

Positive pressure ventilation can be used in remodeled buildings, such as an old stanchion barn converted to a cold calf barn.

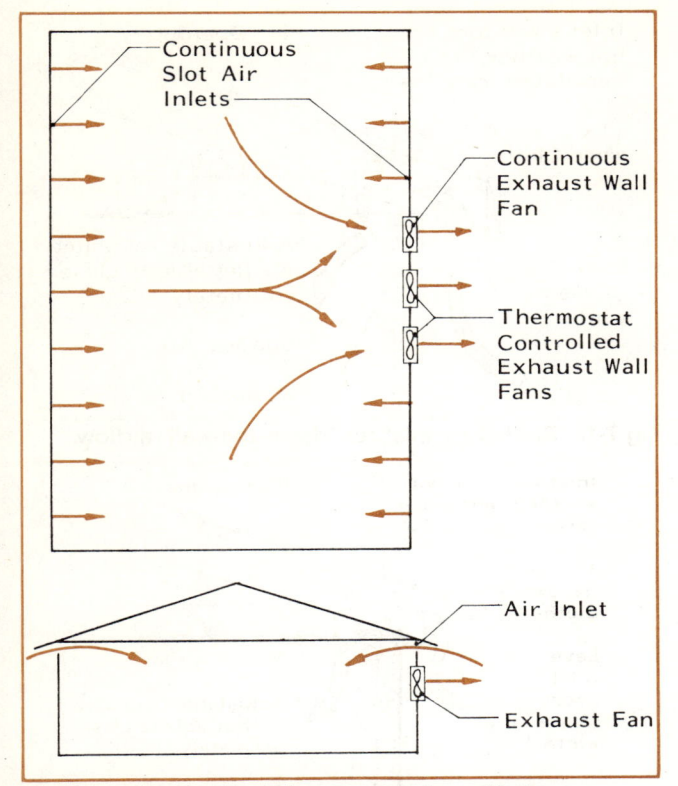

Fig 7-9. Exhaust (negative pressure) ventilating system.
Provide a continuous slot air inlet in both long sidewalls, except over fans and 8' either side during the winter. Group fans in one bank for barns up to 100'-150' long. Use two or more for longer barns.

Provide enough air exchange to maintain inside air temperature near outside air temperatures in the winter. For unweaned calves in individual pens, 10 air changes per hour in winter and 15 air changes per hour for spring and fall are recommended.

A fan in the outside wall forces fresh air through a duct with evenly spaced openings, Fig 7-10, creating a positive indoor pressure. Use a two-speed fan. The duct extends the length of the building to distribute incoming air. Base the distribution duct cross-sectional area on the maximum ventilating rate. Size the duct to have at least 1 ft²/1,000 cfm. Size positive pressure exhaust outlets at 160 in²/1,000 cfm of fan capacity. For example, use one pair of 2" holes for each 40 cfm or one pair of 3" holes per 90 cfm of airflow through the duct.

Ventilating air exhausts through all open doors, windows, and other openings. Provide an exhaust area of at least 1 ft²/600 cfm.

This system is not adequate for summer because of the duct size needed for summer rates. House animals elsewhere in summer or provide additional fans and ducts or separate exhaust-type ventilation.

Negative pressure with recirculation

Negative pressure with recirculating ventilation is not for hot weather ventilation. Use it only for cold or mild weather ventilation. This system can be used to retrofit buildings in which an adequate exhaust system cannot be installed.

A negative pressure with recirculation system includes inlet shutters, a plastic or rigid distribution duct, a duct fan, and exhaust fans, Fig 7-11. In cold and mild weather, ventilating air is distributed through the duct and exhausted through wall fans. A motorized shutter, located in the wall near the duct fan, allows fresh air to enter the duct from the outside or from the attic.

The exhaust fan controls the airflow through the inlet shutters. When the cold weather rate fan is operating, the shutter is opened by a thermostat control. To reduce problems with freeze-up remove the

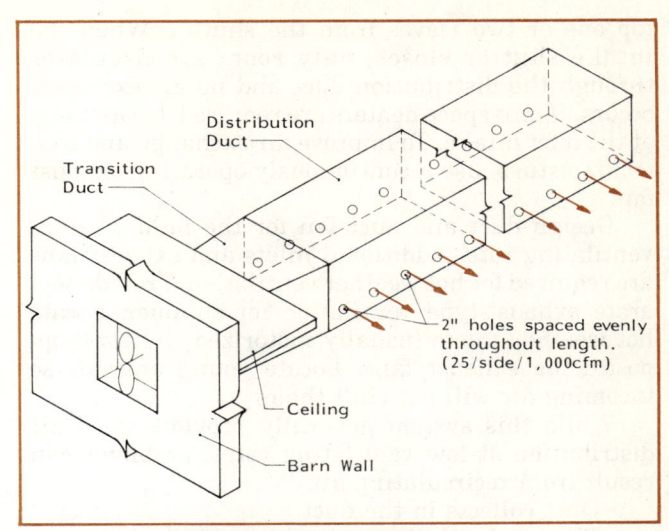

Fig 7-10. Positive pressure ventilation.

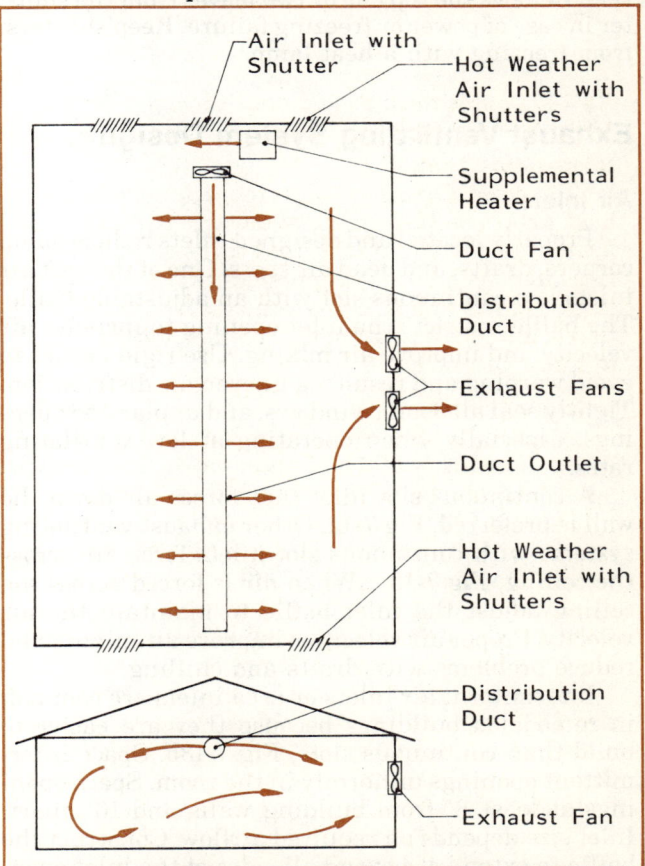

Fig 7-11. Negative pressure with recirculation system.
More than one duct may be needed for buildings over 40' wide. Use continuously operating winter rate exhaust fans to control moisture and maintain fresh air exchange.

top one or two leaves from the shutter. When the intake shutter closes, only room air circulates through the distribution duct and no air exchange occurs. If unit space heaters are required, locate them at the duct intake. To improve air exchange and control moisture, use a continuously operating exhaust fan.

Design duct and duct fan for the mild weather ventilating rate. Additional inlets and exhaust fans are required for hot weather ventilation. Provide separate exhaust-type ventilation for summer. Locate hot weather inlets (usually motorized shutters) opposite the exhaust fans. Locate young animals so incoming air will not chill them.

While this system generally provides good air distribution at low ventilating rates, problems can result from recirculating air.
- Dust collects in the duct.
- The intake shutters can freeze shut.
- Increased air movement can create drafts.
- Use of the same air duct from one animal group to another can transfer dust and bacteria.

The only solution to dust buildup is periodic duct cleaning. This can mean partial disassembly, so consider cleaning ease when constructing homemade ducts. Shutters can freeze shut if no supplemental heat is provided or if the duct fan is too far from the shutters to move warm air past them. Shutter freeze-up is a serious problem because it restricts ventilation. Remove the top one or two leaves from the shutter in case of power or freezing failure. Keep shutters from freezing with a heat lamp.

Exhaust Ventilating System Design

Air Inlets

Properly located and designed inlets reduce damp corners, drafts, and dead air spots. One of the best air inlets is a continuous slot with an adjustable baffle. The baffle restricts the inlet opening to increase air velocity and improve air mixing. Use rigid baffles to avoid warping and resulting uneven air distribution. Tightly seal all doors, windows, and unplanned openings, especially when operating at low ventilating rates.

A continuous slot inlet that forces air down the wall is preferred, Fig 7-12. Other exhaust ventilating systems with continuous slot inlets force air across the ceiling, Fig 7-13a. When air is forced across the ceiling adjust the inlet baffle to maintain the air velocity. Proper air velocities improve air mixing and reduce problems with drafts and chilling.

Intermittent air inlets or area inlets are common in remodeled buildings because they are easier to build than continuous slots, Fig 7-13b. Space intermittent openings uniformly in the room. Space openings at least 8' from building walls and 16' apart. Inlet size depends on required airflow. Construct the baffle to extend 4" beyond all sides of the inlet opening. Do not mount pipes, lights, or other obstructions within 4' of inlets to avoid deflecting cold air down onto animals.

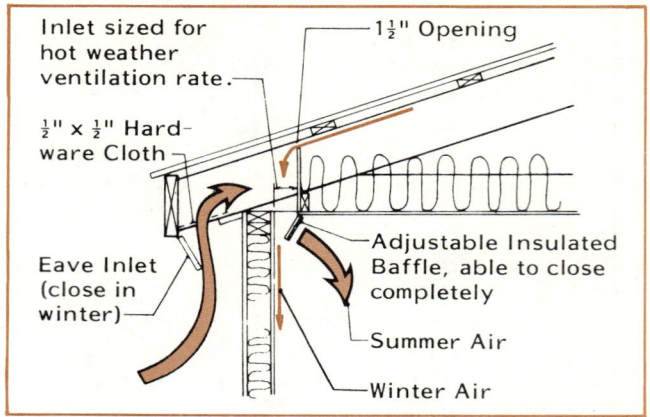

Fig 7-12. Baffled air inlet for "down-the-wall" airflow.

7-13a. Continuous slot air inlet.

7-13b. Intermittent air inlets.

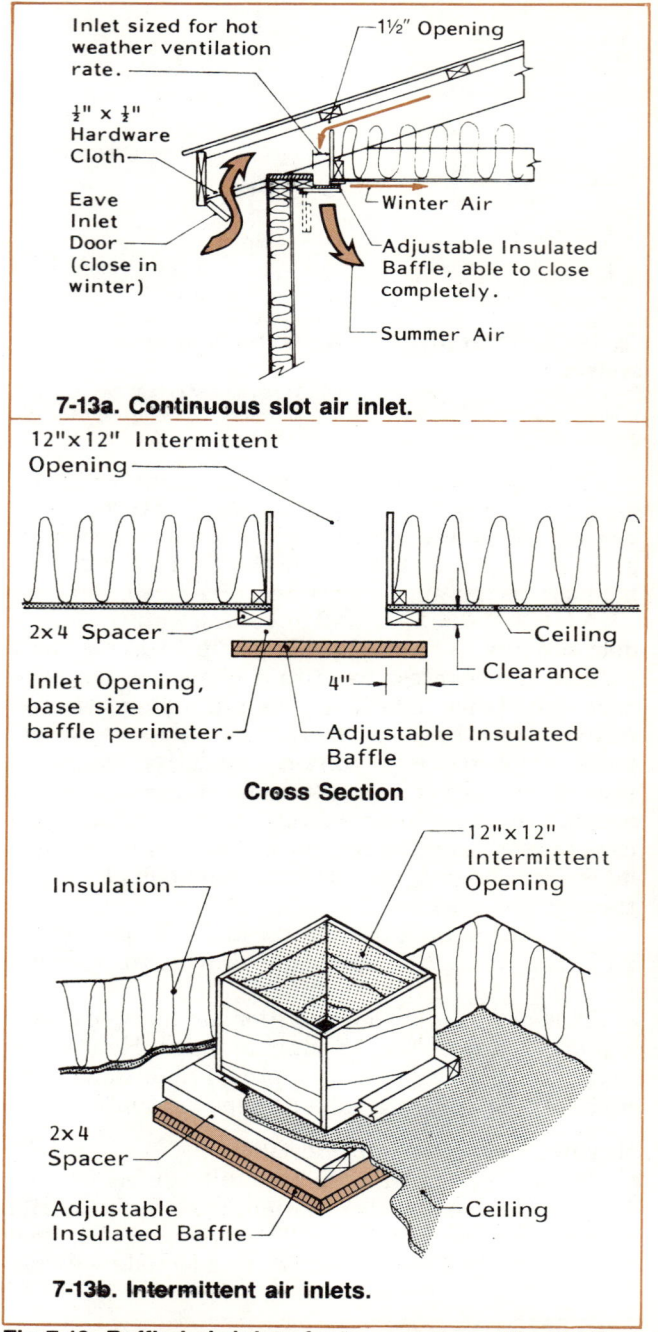

Fig 7-13. Baffled air inlets for "across-the-ceiling" airflow.

Inlet location

For buildings up to 40' wide, place slot inlets at the ceiling along both sidewalls; for wider buildings, add one or more interior ceiling inlets and put fans in both long walls, Fig 7-14.

Locate summer inlets to bring air directly from the outside and not from the attic. Direct outside inlets eliminate the problem of moisture getting into the attic and are often preferred.

During cold weather, close and seal air inlets within 8' of the winter exhaust fans. Locate winter inlets so air enters directly from the outside, from the attic, or both. Adequate attic ventilation and air intakes are needed if air is drawn from the attic during the winter.

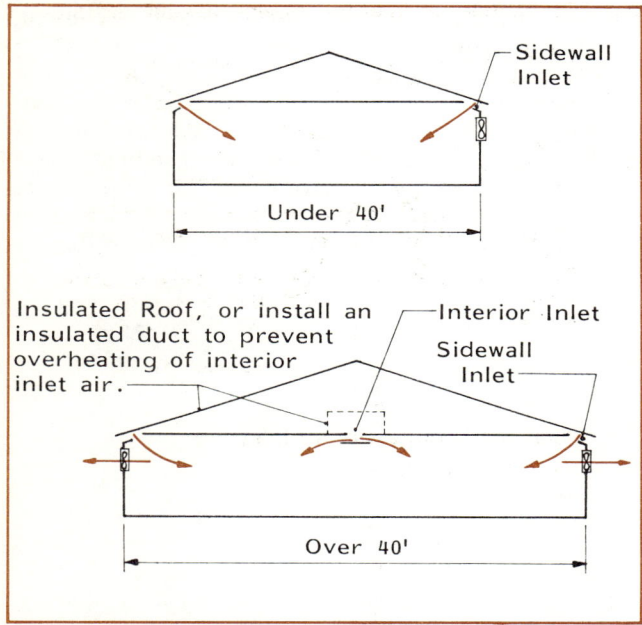

Fig 7-14. Locating slot inlets.

Inlet size

Size inlets to maintain a high velocity (700 to 1,000 fpm) where air leaves the baffle. If the velocity is less, cold air settles too rapidly and can chill the animals; if greater, only the top portion of the room is ventilated and high static pressure decreases fan capacity. Usually, inlets are sized for the hot weather rate, then reduced with a baffle for lower ventilating rates. See the ventilating rates section for desired capacities. Calculate the required slot width using:

Eq 7-1.

$$W = RATE \div (60 \times L)$$

W = slot width, in.
RATE = ventilating rate of fans selected, cfm
L = total slot length, ft

Example 7-1:

Calculate the proper slot openings for a 36'x120', 50-cow stanchion barn. Group exhaust fans in the downwind long walls depending on barn orientation 30' from each end. Locate inlets on both long walls of the barn along the ceiling. For this example, design for 0.04"-0.08" water gauge static pressure with an air velocity of 720 ft/min.

Solution

1. Total slot inlet length is (building length x 2):

    120' x 2 = 240'

    In cold weather, close the inlets over the continuous exhaust fans to prevent short circuited air. **Winter** slot opening is about 200'.

2. Maximum ventilating rate per foot of inlet length is the hot weather rate (Table 7-1) x no. of cows ÷ total slot inlet length.

    470 cfm/cow x 50 cows ÷ 240' = 97.9 cfm/ft

3. A 1" wide slot provides about 60 cfm per foot of length at a velocity of 720 ft/min. The hot weather slot width is:

    97.9 cfm/ft ÷ 60 cfm/ft-in = 1.63"

4. In cold weather, the ventilating rate is much less. The required rate is the cold weather rate (Table 7-1) x no. of cows ÷ total slot inlet length. Use the winter slot opening of 200'.

    50 cfm/cow x 50 cows ÷ 200' = 12.5 cfm/ft

5. To provide the cold weather rate, partly close the slot with a baffle. The baffled slot width is:

    12.5 cfm/ft ÷ 60 cfm/ft-in = 0.21"

Inlet control

Air inlet control is critical to good ventilation. The slot inlet size must change with ventilating rate changes. Operate manually-adjusted baffles from one location with a winch and cable system. Install a manometer next to the winch for more accurate baffle adjustments. A manometer measures static pressure, which is the difference in atmospheric pressure inside and outside a building. A slot width sized to deliver air at 700 to 1,000 fpm creates a static pressure of 0.04"-0.06" water column. Keep doors and windows closed when the negative pressure ventilating system is running or air may not be drawn in through the slot inlets.

Fans

Fan selection guidelines

Select fans to move enough air against at least ⅛" static pressure. Variable-speed fans have poor pressure ratings at low speeds and may not deliver enough air when operating against wind. Provide the cold weather ventilation capacity with a single-speed or two-speed fan. Provide additional capacity with single-speed, multiple-speed, and/or variable-speed fans. Purchase fans that have an Air Movement and Control Association (AMCA) "Certified Rating" seal or equivalent testing and rating.

Use fans designed specifically for animal housing. Buy totally enclosed, split phase, or capacitor-type farm duty fan motors. Wire each fan to a separate circuit to avoid shutdown of the entire ventilating system if one motor blows a fuse. Protect each fan with a time-delay fused switch at the fan. Size time-delay fuses at 25% over fan amperage.

Select fan motors with thermal overload protection and manual reset switches. Manual reset switches reduce the safety hazard of a fan starting while being checked and reduce on-off cycling that can damage the motor.

Fan location

With a rigid baffle slot inlet, fan location has little effect on air distribution. Place winter fans in the downwind sidewall for buildings less than 40′ wide and in both sidewalls for wider buildings. To prevent air from becoming too stale, place fans no more than 75′ from the farthest inlet.

Put anti-backdraft shutters on all noncontinuous fans. Place shutters on the inside (animal side) of fans to decrease freezing problems. Space mechanically ventilated buildings at least 35′ apart, so fans do not blow foul air into the intakes of adjacent buildings.

Fan controls

Accurate sensing and control devices make a ventilating system nearly automatic and regulate the flow of air required to maintain desired conditions. Seasonal management and maintenance are still required. A line thermostat is the most common and basic control; it has a temperature-sensing element and a switch. Humidistats, which have an element sensitive to air moisture content, are not recommended for livestock buildings because of unreliability of currently available equipment in livestock environments.

Choose thermostats designed especially for livestock housing with an on-off range of 5 F or less. Locate the thermostat:
- At or near the center of the building width.
- Out of animal reach.
- Away from cold walls and ceilings.
- About 4′-5′ off the floor.
- Out of the path of furnace exhausts, inlet air, and direct sunlight.

Determining fan sizes and thermostat settings

Select fan capacities and thermostat settings for a **gradual** airflow increase as outside temperatures increase from cold to mild. Larger airflow increases are common between mild and hot weather. Use several single-speed fans or a combination of single speed, multi-speed fans, or variable-speed fans. Multiply the recommended ventilating rates in Table 7-1 by the number of animals, then select fan sizes from product literature.

Small and large fans are needed to exhaust different amounts of air under varying weather conditions. A duct around the small fan permits it to exhaust cooler air from about 15″ above the floor during winter to remove moisture and conserve heat, Fig 7-15.

Most thermostats have an on-off range of about 3-5 F, and fans are normally staged to come on in 2-4 F increments.

Example 7-2:

Calculate the cold, mild, and hot weather ventilating rates for a 20-head calf barn. Table 7-1 gives the following ventilating rates:

Cold weather: 15 cfm/calf x 20 calves = 300 cfm total

Mild weather: 35 cfm/calf x 20 calves = 700 cfm (Total = 300 + 700 = 1,000 cfm)

Hot weather: 50 cfm/calf x 20 calves = 1,000 cfm (Total = 1,000 + 1,000 = 2,000 cfm)

The desired room temperature for stanchion and calf barns is 45-55 F. Set heater thermostats in calf barns about 2 F below the desired minimum room temperature (43 F). With a 5 F range on the furnace thermostat, the furnace shuts off at 48 F. The cold weather fan runs continuously so it has only a safety thermostat. Set the thermostat of the smallest mild weather fan at about 3 F above the furnace-off temperature (51 F) or 3 F above desired temperature (48 F) if no heater is used. Make sure the heater does not operate after mild weather fans turn on. Set the other warm weather fans to turn on at increasing increments of 2-4 F (52 F, 56 F, etc.). The more fans in the room, the smaller the increments.

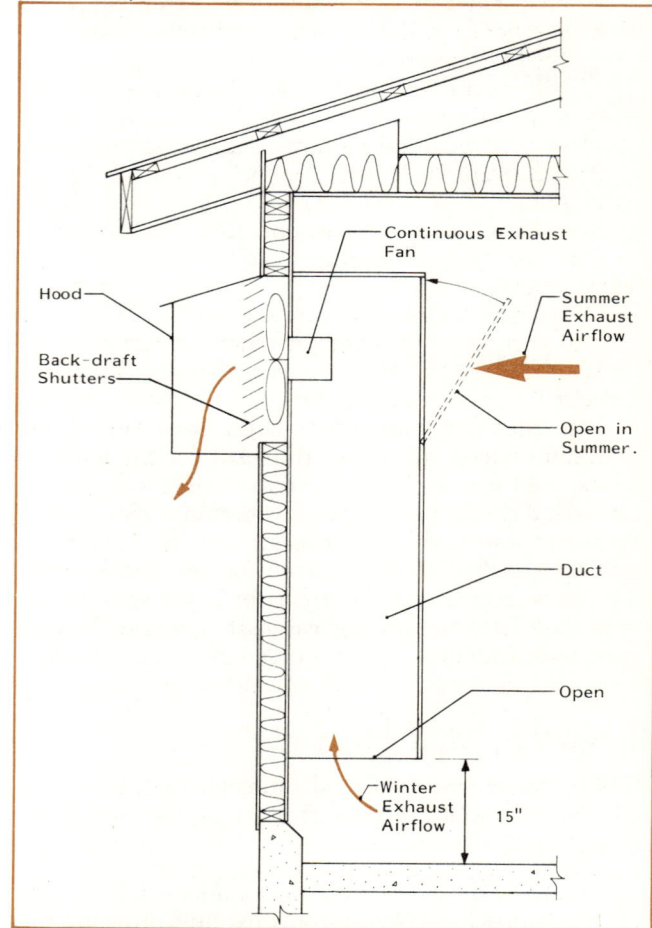

Fig 7-15. Heat conservation duct for continuous exhaust fan.

7.10

Environmental Control System Maintenance

Environmental control systems require periodic, conscientious, and thorough maintenance. Use the following to develop a maintenance schedule:

Every month:
- Clean fan blades and shutters. Dirty fan shutters can decrease fan efficiency by 25%. Shut off power to thermostatically controlled fans before servicing them.
- Check fans with belt drives for proper tension and correct alignment. If too tight, belts can cause excessive bearing wear; if too loose, excessive slippage and belt wear result.
- During the heating season, remove dust from heater cooling fins and filters, and check gas jets and safety shut-off valves for proper operation.
- Test emergency ventilation.
- Make certain that shutters open and close freely. Apply a few drops of graphite (not oil or grease) to fan shutter hinges.
- Clean motors and controls. Dirty thermostats do not sense temperature changes accurately or rapidly. Dust insulates fan motors and prevents proper cooling. If dust is allowed to build up, the motor can overheat.
- Check, clean, and adjust slot air inlet openings.

Every 6 months:
- Most ventilating fans have sealed bearings and do not require lubrication. However, oil other bearings with a few drops of SAE No. 10 non-detergent oil or as recommended by the manufacturer. Do not overlubricate.

Every year:
- Clean and repaint chipped spots on fan housings and shutters to prevent further corrosion.
- Each fall, turn off and cover summer fans (not cold or mild weather fans) with plastic or an insulated panel on the warm (animal) side of the fan and disconnect the power supply. Uncover in spring.
- Check slot air inlets for debris and baffle condition.
- Check gable and soffit air inlets for blockages.
- Check plastic baffle curtains and distribution ducts. They become brittle with age and require replacement.
- Check recirculation air ducts for excess dust.

Emergency Ventilation

Provide for emergency ventilation in all mechanically ventilated buildings because of the danger of animal suffocation. Consider simple, manually-opened sidewall doors as well as a sophisticated electric generator system that automatically runs fans if the power fails. Magnet-locked ventilating doors, which drop open when electrical power is cut off or room temperature rises sharply, are available. Consider installing an alarm system to warn you when electrical power is off. Test your emergency systems monthly or according to manufacturers' instructions.

Manure Pit Ventilation

Pit ventilation reduces manure gas in the animal area, reduces odor levels, improves air distribution, and helps warm and dry floors. All warm, slotted floor buildings can benefit from pit ventilation.

Allow at least 12" clearance between the bottom of slat support beams and the manure surface. Variable-speed fans are not recommended. Fans with a corrosion resistant finish are required.

Annexes

An annex is common in totally slotted buildings, Fig 7-16. It does not provide air distribution through the floor as uniformly as a duct, but it is less expensive.

Design annex pit ventilation to supply at least the cold weather rate but no more than the mild weather rate. Locate annexes so no point in the pit is farther than 50′ from an annex.

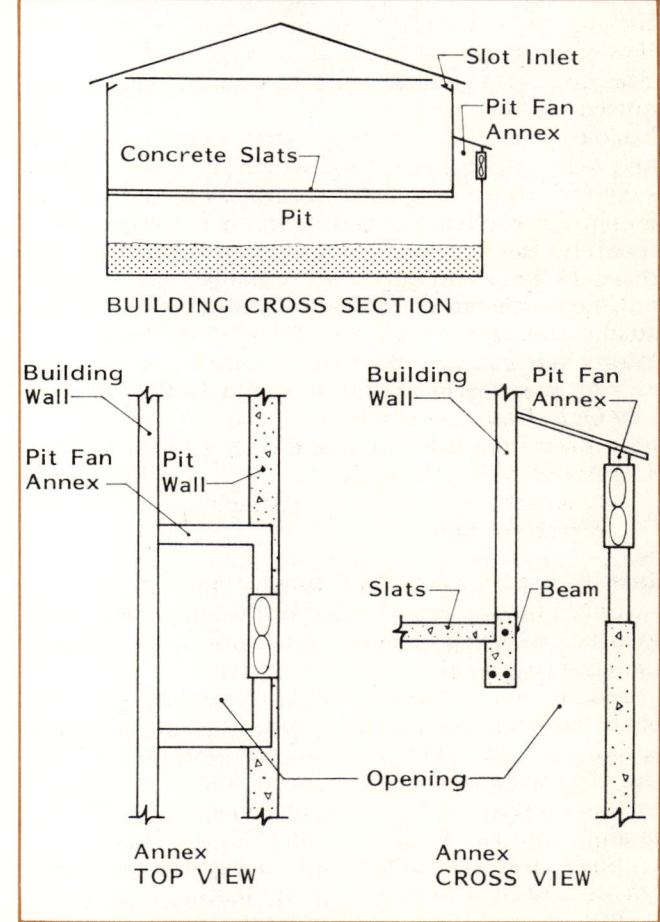

Fig 7-16. Exhaust pit fan annex.
Size pit openings to supply the cold or mild weather rate. Provide an air velocity of 800 ft/min through the opening.

Pit Fan Installation

Insulate and use corrosion resistant construction for pit fan housings. To prevent short-circuiting between fans, put each pit fan in its own duct or annex. Shutters are not necessary on continuous operation fans.

Attic Ventilation

Provide for continuous ventilation of an attic space to minimize moisture buildup caused by weather changes. Use gable louvers and ridge vents. Provide 1 in² of net vent area for each ft² of ceiling area.

Heaters

Supplemental heat may be required in some buildings, e.g. calf and treatment barns, to maintain the desired temperature. Size the heating system to equal the heat loss from the building through conduction and ventilating air. To estimate heater size for warm calf barns, use:

Eq 7-2.

HEAT = CFM x TD x 2.0
HEAT = heater size, Btu/hr output
CFM = ventilating rate, cfm
TD = Temperature difference, Ti – To, F
Ti = desired inside temperature, F
To = outside winter design temperature, F

Unit Space Heater

Unit space heaters heat room air directly. Unfortunately, they recirculate dusty, wet, corrosive air through the furnace, often requiring high maintenance. Clean and lubricate unit heaters once a month during the heating season and more often in very dusty environments. If possible, place heaters to blow along the coldest wall to reduce cold drafts and radiant heat losses to the wall. With more than one heater, arrange them to create a circular air pattern within the room.

Air Make-up Heaters

Air make-up heaters are unvented units that heat only the incoming ventilating air. They require less service than units that recirculate room air. Fuel flow to the burner is regulated to maintain a constant exit air temperature.

Air make-up heaters supply ventilating air, so close down air inlets accordingly. Size air make-up heaters so they provide no more air than the cold weather exhaust fan.

Air make-up heaters exhaust the products of combustion into the building which makes them more efficient than vented heaters (about 90% vs. 70%). However, water vapor is one of the gases produced (about 1 lb water/lb propane burned), so the ventilating rate must be sufficient to remove this extra moisture. Provide 4 cfm extra ventilation for each 1,000 Btu/hr of heater capacity. The added efficiency of unvented heaters is canceled out by the increased ventilation requirements. This makes the operation cost of vented and unvented heaters about the same.

Solar Heat

The limited heating needs in dairy buildings make solar energy expensive.

Reusing Exhaust Air

Reusing exhaust air from one livestock building to ventilate an adjacent one is not recommended. Exhaust air carries moisture, gases, dust, and pathogenic organisms which can cause problems that outweigh the value of the energy savings.

Heat Exchangers

Heat exchangers extract heat from exhaust air that otherwise would be lost. Although this concept has merit, dust and condensation cause equipment problems and decreased efficiency. Heat exchangers may have some use in calf housing. Consider the cost of operating extra and larger fans.

Insulation

Insulation is any material that reduces heat transfer from one area to another. Although all building materials have some insulation value, the term "insulation" usually refers to materials with a relatively high resistance to heat flow. The resistance of a material to heat flow is indicated by its **R-value**. Good insulators have high R-values. See Table 7-2.

Insulation Types

Batts and blankets are available in 1"-8" thicknesses and in widths to fit 16", 24", and 48" stud spaces. Batts are 4'-8' long, and blankets are up to 100' long. Materials are fiberglass, mineral wool, or cellulose fibers. Some batts or blankets have a paper or aluminum face to serve as a partial vapor barrier. An additional plastic vapor barrier is required in dairy buildings.

Loose-fill insulation is packaged in bags and can be mineral wool, cellulose fiber, vermiculite, granulated cork, and/or polystyrene. It is easy to pour or blow above ceilings and in walls. Poor quality insulation can settle in walls, leaving the top inadequately insulated.

Rigid insulation is made from cellulose fiber, fiberglass, polystyrene, polyurethane, or foam glass and is available in ½"-2" thick by 4' wide panels. Some types have aluminum foil or other vapor barriers attached to one or both faces. Rigid insulation can be used for roofs and walls or as a ceiling liner. It can also be used along foundations (perimeter insulation) or buried under concrete floors (if waterproof and protected from physical and rodent damage).

Support rigid insulation at least 2' o.c. Use tongue-and-groove panels or seal the joints with caulk or tape to prevent moisture from passing through the joints. Check for flammability and toxic gas production if burned. Check if your insurance company requires rigid insulation to be protected with fire-resistant materials.

Foamed-in-place insulation is usually obtained only through commercial applicators because it requires special equipment and experienced workers. Substandard application can cause excessive shrinkage. Formaldehyde-based insulation is not recommended because gases may be offensive or toxic. Apply a separate vapor barrier.

Table 7-2. Insulation values.
Adapted from *ASHRAE Handbook of Fundamentals*. Values do not include surface conditions unless noted otherwise. All values are approximate.

Material	R-value Per inch (approximate)	R-value For thickness listed
Batt and blanket insulation		
Glass or mineral wool, fiberglass	3.00-3.80[a]	
Fill-type insulation		
Cellulose	3.13-3.70	
Glass or mineral wool	2.50-3.00	
Vermiculite	2.20	
Shavings or sawdust	2.22	
Hay or straw, 20″		30+
Rigid insulation		
Expanded polystyrene,		
Extruded, plain	5.00	
Molded beads, 1 pcf	5.00	
Molded beads, over 1 pcf	4.20	
Expanded rubber	4.55	
Expanded polyurethane, aged	6.25	
Glass fiber	4.00	
Wood or cane fiberboard	2.50	
Polyisocyanurate	7.04[b]	
Foamed-in-place insulation		
Polyurethane	6.00	
Urea formaldehyde	4.00	
Building materials		
Concrete, solid	0.08	
Concrete block, 3 hole, 8″		1.11
Lightweight aggregate, 8″		2.00
Lightweight, cores insulated		5.03
Metal siding	0.00	
Hollow-backed		0.61
Insulated-backed, ⅜″		1.82
Lumber, fir and pine	1.25	
Plywood, ⅜″	1.25	0.47
Plywood, ½″	1.25	0.62
Particleboard, medium density	1.06	
Hardboard, tempered, ¼″	1.00	0.25
Insulating sheathing, 25/32″		2.06
Gypsum or plasterboard, ½″		0.45
Wood siding, lapped, ½″x8″		0.81
Windows (includes surface conditions)		
Single glazed		0.91
With storm windows		2.00
Insulating glass, ¼″ air space		
Double pane		1.69
Triple pane		2.56
Doors (exterior, includes surface conditions)		
Wood, solid core, 1¾″		3.03
Metal, urethane core, 1¾″		2.50
Metal, polystyrene core, 1¾″		2.13
Floor perimeter (per ft of exterior wall length)		
Concrete, no perimeter insulation		1.23
With 2″x24″ perimeter insulation		2.22
Air space (¾″-4″)		0.90
Surface conditions		
Inside surface		0.68
Outside surface		0.17

[a]The R-value of fiberglass varies with batt thickness. Check package label.
[b]Time aged value for board stock with gas barrier quality aluminum foil facers on two major surfaces.

Sprayed-on insulation applied to inside or outside surfaces is difficult to protect with an adequate vapor barrier. Exterior application must also be protected from sunlight. Improperly installed insulation may peel off.

Selecting Insulation

Consider the following factors:
- **Ease of installation.** Will you do it yourself? Must siding or inner wall surface be removed? Some materials are harder to handle or take more time to install, which increases labor. Others are irritating to eyes and skin, requiring protective clothing and masks.
- **Thickness.** A ceiling may allow for many inches of material; a wall or roof may have thickness limitations.
- **Fire resistance.** Is a fire-resistant liner required to prevent rapid flame spread? Check with your insurance company before choosing insulation.
- **Animal contact.** Will the insulation be exposed to physical damage, requiring a protective covering that increases the cost?
- **Cost.** What will the different types of insulation cost considering R-value, preparation, installation, protection, and purchase price? Material and labor costs can vary significantly.

Insulation Levels

Minimum insulation levels for warm barns are based on winter heat loss. Provide a minimum insulation level of R=11 in the walls and R=19 in the ceiling of warm barns. Additional insulation can reduce condensation. See Table 7-3 for suggested insulation levels based on winter degree days.

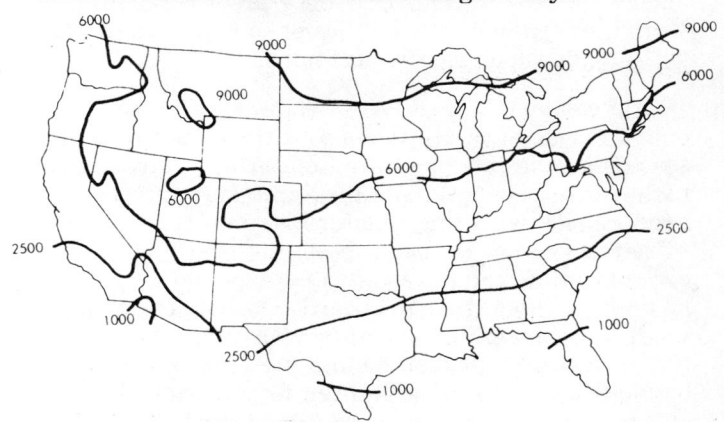

Fig 7-17. Winter degree days.
Accumulated difference between 65 F and average daily temperature for all days in the heating season.

Table 7-3. Minimum dairy insulation levels.
R-values are for building sections. In cold barns with mature animals, no insulation is needed in the walls or ceiling, but a total R-value of 3 or more can be used in the ceiling to control roof condensation and frosting and reduce summer heat load.

Winter degree days	Minimum R-values, warm barn Walls	Ceiling	Perimeter
2,500 or less	11	22	4
2,501-6,000	11	25	6
6,001 or more	16	33	8

Insulation is not needed in cold barns with mature animals. However, adding insulation to the roof can control roof condensation and frosting and reduce summer heat load. Use a total roof section R-value of 3 or more.

Installing Insulation

Figs 7-18 to 7-21 show common construction methods for insulated roofs, ceilings, walls, and foundations. Cracks around window and door frames, pipes, and wires result in cold spots, drafty areas, or condensation, and reduce ventilating inlet effectiveness. Caulk cracks and joints on outside surfaces.

Perimeter insulation reduces heat loss through the foundation and eliminates cold, wet floors. Insulate concrete foundations by covering the foundation exterior to a minimum of 24" below the ground line for thermal protection and 36" for rodent control. Rodents can burrow up to 36" below grade damaging insulation and foundations. Use 2" rigid insulation and protect it with asbestos-cement board or fiberglass board above and below ground.

Maximizing Insulation Effectiveness

Vapor barriers

Wet insulation increases heat loss and building deterioration. Vapor barriers are important in restricting moisture migration through walls, ceilings, and roofs. Vapor barriers are classified by a permeability rating measured in perms.

Install 4 to 6 mil polyethylene (plastic) vapor barriers on the warm side of all insulated walls, ceilings, and roofs. Use polyethylene vapor barriers underneath concrete floors and foundations to control soil moisture penetration. Use waterproof rigid insulation if in contact with soil. Refer to Figs 7-18 to 7-21 for proper vapor barrier locations.

Fire protection

The rate at which fire moves through a room depends on the interior lining material. Many plastic foam insulations have high flame spread rates. To reduce risk with these materials, protect them with fire-resistant coatings such as:
- ½" thick cement plaster.
- Fire rated gypsum board. Do not use in high moisture environments such as animal housing.
- ¼" thick sprayed-on magnesium oxychloride (60 lb/ft^3) or ½" of lighter, foam material.
- Mineral asbestos board ⅛"-⅜" thick.
- Fire rated ½" thick exterior plywood.

Doors and windows

Locate doors on the downwind side and use insulated and weatherstripped entry doors. For upwind entries, a vestibule or hallway between outer and inner doors prevents cold wind from blowing directly into the building. To conserve floor area, consider making the vestibule entrance a part of an office, storage area, wash room, or hallway.

Minimize windows and skylights in animal housing. In warm buildings, windows and skylights increase winter heat loss. The insulation value of windows (R = 1-2.5) is well below the R-value of an insulated wall (R = 13-15). In cold barns, skylights can cause roof leaks and increase the summer heat load.

If remodeling a building, consider replacing all windows with insulated removable panels or permanent wall sections. Hinged or removable panels can be opened for summer ventilation.

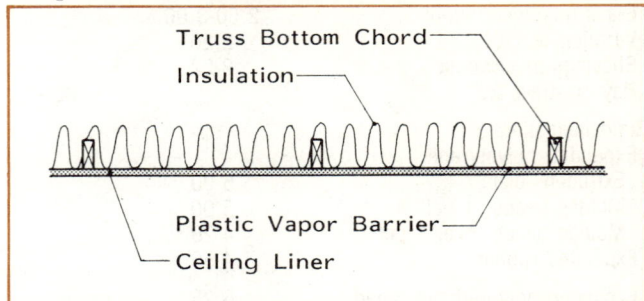

Fig 7-18. Insulating ceilings.
Loose-fill, batt, or blanket insulation recommended for insulating ceilings in environmentally controlled buildings. Insulation must cover the truss bottom chord.

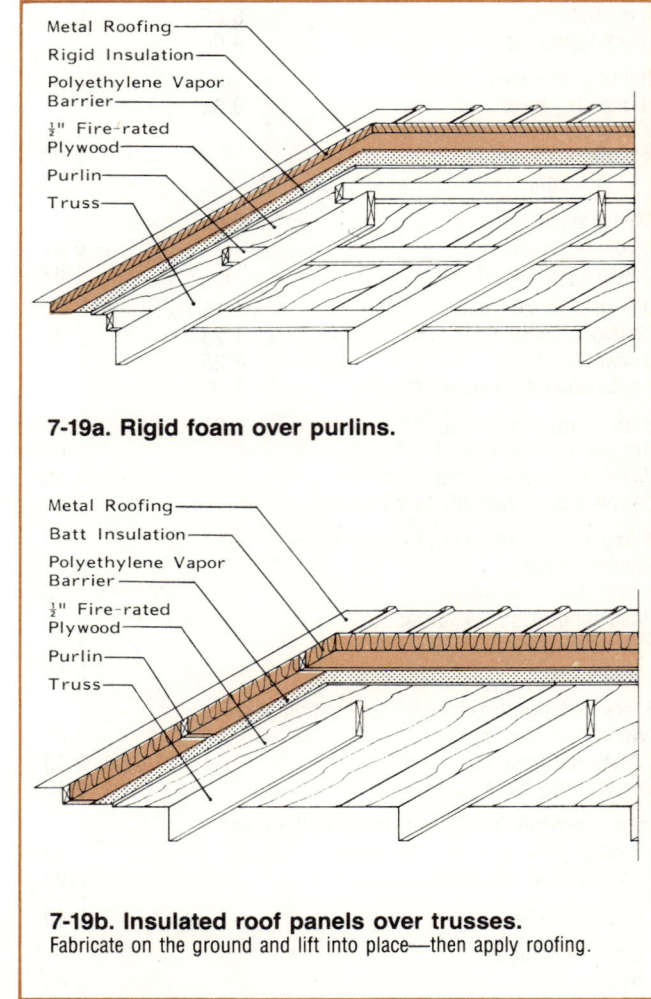

7-19a. Rigid foam over purlins.

7-19b. Insulated roof panels over trusses.
Fabricate on the ground and lift into place—then apply roofing.

Fig 7-19. Insulating roofs.

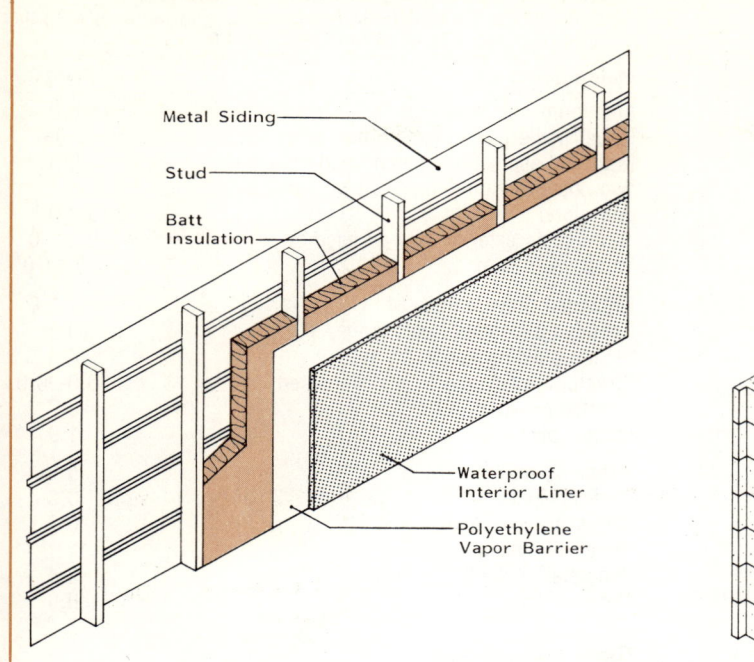

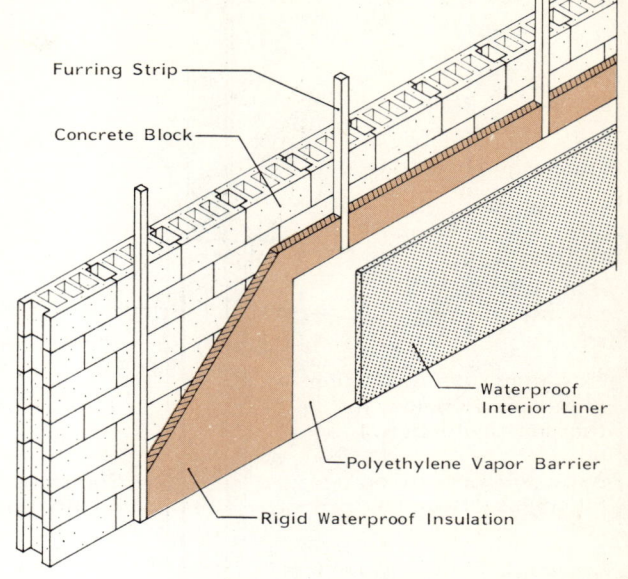

7-20a. Stud wall insulation.
Approximate R = 12 if 2x4 studs, 20 if 2x6 studs. Common in warm buildings.

7-20c. Concrete block wall insulation.
Approximate R = 10 if standard blocks, 14 if lightweight blocks with cores filled. Can be used for remodeling.

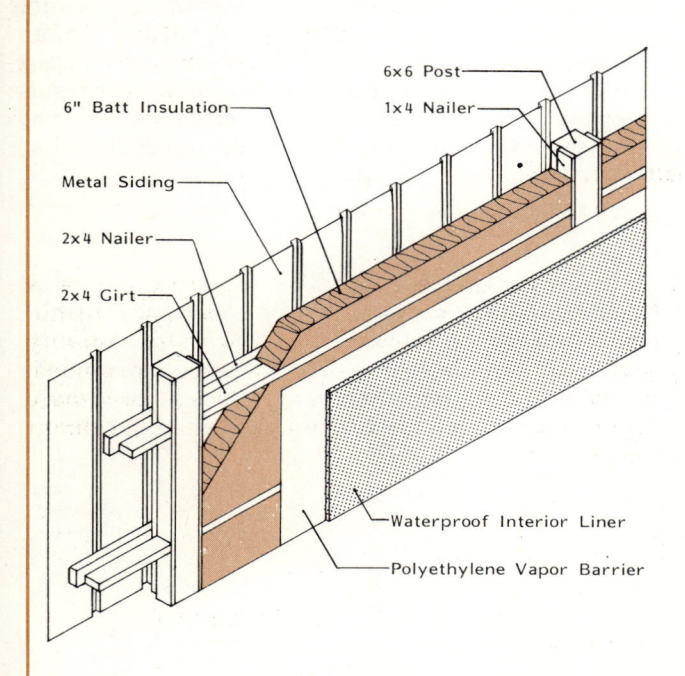

7-20b. Post frame building wall with 6″ batt insulation.
Approximate R = 21.

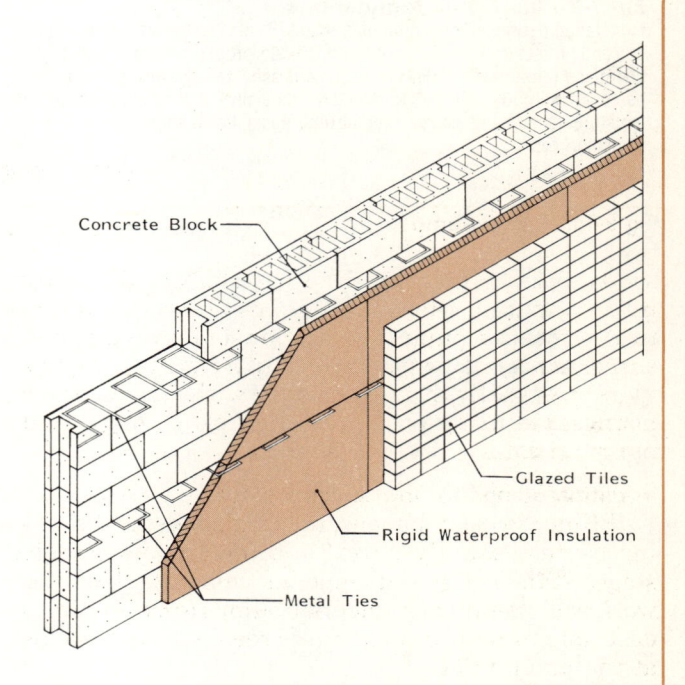

7-20d. Glazed tile and block wall insulation.

Fig 7-20. Insulating walls.

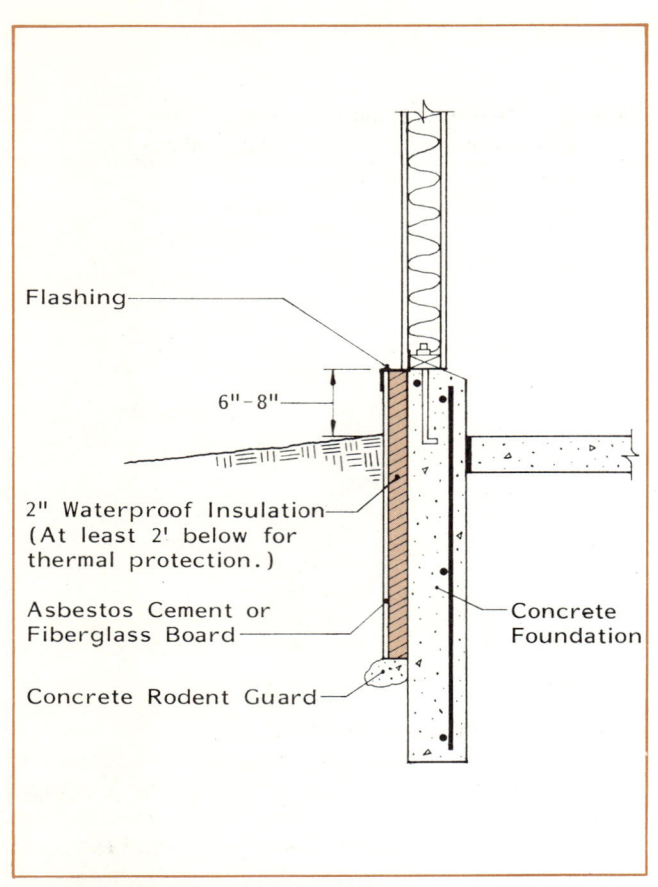

Fig 7-21. Insulating foundations.
Foundation perimeter insulation on outside (R = 2.2). Use waterproof insulation and protect from damage with a rigid, waterproof covering—high density fiberglass reinforced plastic or ¼" cement asbestos board preferred. Tempered ¼" hardboard or ⅜" foundation grade plywood resist physical and moisture damage but are not rodentproof. Backfill with soil to within 6"-8" of insulation top.

Table 7-4. Permeability of building materials.
Adapted from the *1981 ASHRAE Handbook of Fundamentals*. A vapor barrier should have a perm rating of less than 1.0.

Material	Perms
Aluminum foil, 1-mil	0.0
Polyethylene plastic film, 6-mil	0.06
Kraft and asphalt laminated building paper	0.3
Two coats of aluminum paint (in varnish) on wood	0.3-0.5
Three coats exterior lead-oil base on wood	0.3-1.0
Three coats latex	5.5-11.0
Expanded polyurethane, 1"	0.4-1.6
Extruded expanded polystyrene, 1"	0.6
Tar felt building paper, 15 lb	4.0
Structural insulating board, uncoated, ½"	50.0-90.0
Exterior plywood, ¼"	0.7
Interior plywood, ¼"	1.9
Tempered hardboard, ⅛"	5.0
Brick masonry, 4"	0.8
Cast-in-place concrete wall, 4"	0.8
Glazed tile masonry, 4"	0.12
Concrete block, 8"	2.4
Metal roofing	0.0

Birds and rodents

To prevent rodent damage, cover exposed perimeter insulation with a protective liner and maintain a rodent control program.

To prevent bird damage, cover exposed insulation with a protective liner and construct buildings so birds cannot roost near the insulation. An aluminum foil covering is not sufficient protection. Consider screening all vent openings. Use ½" hardware cloth for air intake openings and ¾" hardware cloth for air outlet openings. Screened vent outlets are very susceptible to freezing shut in cold climates. Ice may have to be regularly knocked off during cold spells.

8. MANURE MANAGEMENT

A complete manure management system is necessary to a dairy operation. Goals are to:
- Maintain good animal health through sanitary facilities.
- Minimize pollution of air and water.
- Reduce odors and dust.
- Control insect reproduction.
- Comply with local, state, and federal regulations.
- Balance capital investment, cash flow requirements, labor, and nutrient use.

Manure can be handled as a solid, a semi-solid, or liquid. The amount of bedding or dilution water influences the form. In turn, the form influences the selection of collection and spreading equipment and the choice of storage type.

Handle solid and semi-solid manure with tractor scrapers, front-end loaders, or mechanical scrapers. Conventional spreaders are common for land application.

Handle liquid manure with scrapers, flushing systems, gravity flow gutters, or storage below slotted floors. Spread liquids on fields with tank wagons or irrigation equipment.

Waste Volumes

Milking Center Effluent

Management and equipment greatly affect the volume of effluent from milking and cleaning operations. If cow udders are washed with paper towels and disinfectant, there is little discharge water. An automatic pre-wash stall can use 9 gal/day per cow washed. Similarly, floors can be washed with relatively little water and a stiff-bristle broom or hosed down.

Milking center effluent includes:
- Dilute liquid manure with feed, bedding, and hoof dirt. The solids may settle or float.
- Dilute milk and cleaners from equipment washing. The suspended solids do not settle easily and residual cleaning chemical concentrations can affect the treatment and disposal method.
- Concentrated milk products containing colostrum, medicated, or spilled milk.
- Clean water from final pipeline rinses and a water-cooled condensor.

Milk solids include fat, albumin, and lactose. These solids do not settle out, can cause severe odors in anaerobic lagoons, and can plug absorption fields and dry wells. Dispose of milk solids by field spreading or aerobic lagoon treatment.

Table 8-1 has data from manufacturers' recommendations and field observations. Small operations tend to use less water, but more per day per cow, for cleaning floors and equipment. A 100-cow operation with automatic washing equipment can use over 800 gal/day or 3,200 ft^3/month.

Consult local health and regulatory authorities and have all plans approved before constructing any manure handling system.

Table 8-1. Volume of milking center effluent.

Washing operation	Water volume
Bulk tank	
Automatic	50-60 gal/wash
Manual	30-40 gal/wash
Pipeline	
in parlor	75-125 gal/wash
(Volume increases for long lines in a large stanchion barn.)	
Pail milkers	30-40 gal/wash
Misc. equipment	30 gal/day
Cow prep	
Automatic	1-4½ gal/washed cow
Estimated average	2 gal/washed cow
Manual	¼-½ gal/washed cow
Parlor floor	40-75 gal/day
Milkhouse floor	10-20 gal/day
Toilet	5 gal/flush

Septic tank with absorption field

This system is not recommended. Manure and milk solids plug the soil absorption field. A 200 gal septic tank providing a 2-day detention time can handle toilet room wastes. Use absorption fields only in well drained soils.

Settling tank with surface disposal

A settling tank removes solids that float or settle. Removing solids reduces problems with pumping, distribution, and treatment. A compartmented tank or two or more tanks in series improve solids removal.

Design settling tanks to handle at least five times the daily water volume but not less than 2,000 gal. Construct the tank with a length to width ratio between 3:2 and 5:1 to aid settling. Pump out the tank when ¾ full of solids.

Irrigation or a vegetative infiltration area disposes of settling tank effluent. See the manure handling section for irrigation and vegetative infiltration area details.

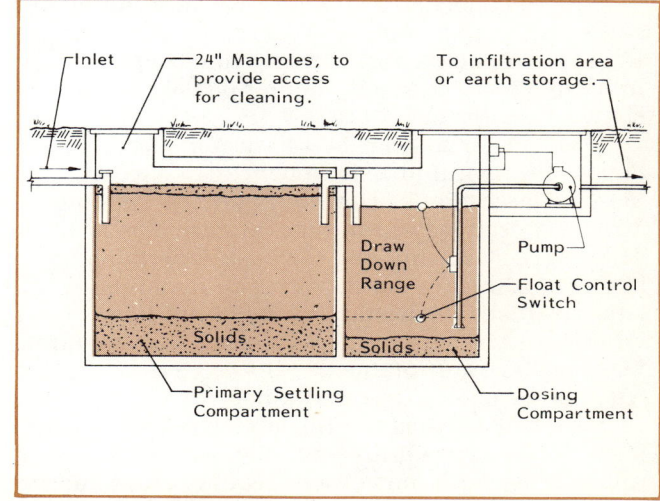

Fig 8-1. Settling tank.

Lagoons

Large herds with pre-wash stalls produce too much water for underground absorption systems. Lagoons or other liquid manure systems provide practical alternatives. A well managed aerobic lagoon effectively decomposes all milking center effluent in moderate climates. Provide an area of 50 to 60 ft²/cow with a depth of 3'-5'.

Other methods

Holding tanks provide short-term storage until conditions allow spreading on nearby cropland, pastures, or woodlots. Size the storage volume to hold at least 15 days of discharge.

Manure

Dairy cattle produce about 1.32 ft³ (62 lb) of manure per day per 1,000 lb of liveweight. Fresh manure, a mixture of urine and feces, is about 12% solids, wet basis. Actual solids content depends on amounts of bedding or extra liquid added and drying. Table 8-2 gives more exact production values.

Table 8-2. Dairy manure production.
Manure at 87.3% water and 62 lb/ft³.

Animal size, lb	Total manure production lb/day	ft³/day	gal/day	N lb/day	P lb/day	K lb/day
150	12	0.19	1.5	0.06	0.010	0.04
250	20	0.33	2.4	0.10	0.020	0.07
500	41	0.66	5.0	0.20	0.036	0.14
1,000	82	1.32	9.9	0.41	0.073	0.27
1,400	115	1.85	13.9	0.57	0.102	0.38

Bedding

Bedded housing, such as loafing barns, maternity pens, and calf barns, requires handling manure as a solid. Adding bedding materials to manure increases the solids content. Usually about 12 lb of bedding per 100 lb of fresh manure is needed to handle dairy manure as a solid.

Handled properly, almost any kind of fibrous material can be effective dairy cow bedding. But some types of manure handling equipment and storages restrict the use of certain bedding materials.

Long or chopped straw works well for bedding free stalls and can be handled with tractor scrapers and solid piston pumps. But use only fine-chopped straw in limited quantities with hollow piston pumps or in slotted floor barns.

Wood materials can be satisfactory for bedding. Sawdust from kiln-dried lumber can be successful on hard-surfaced stalls. However, as deep bedding on dirt surfaces, it may encourage coliform mastitis. The same is true with mill-run sawdust. A combination of 60%-70% shredded bark, with the remainder mill-run sawdust, is satisfactory bedding, keeps cows cleaner, moves through slots, and causes few liquid manure handling problems. Do not use woodchips when other choices are available.

Sand is not a good bedding material with manure storages. To successfully load manure out of a semi-solid storage with conventional equipment, some dry material with water absorption capabilities must be included. Sand does not absorb water. In a liquid storage, sand settles to the bottom and tends to clog pipes and increase equipment wear. The same is true for crushed limestone, paper mill sludge, clay, and other materials that are impervious to water.

Table 8-3. Bedding materials.
Approximate water absorption of dry bedding (typically 10% moisture).

Material	lb water/lb bedding
Wood	
Tanning bark	4.0
Dry fine bark	2.5
Pine chips	3.0
Sawdust	2.5
Shavings	2.0
Needles	1.0
Hardwood chips, shavings or sawdust	1.5
Corn	
Shredded stover	2.5
Ground cobs	2.1
Straw	
Flax	2.6
Oats	
Combined	2.5
Chopped	2.4
Wheat	
Combined	2.2
Chopped	2.1
Hay, chopped mature	3.0
Shells, hulls	
Cocoa	2.7
Peanut, cottonseed	2.5
Oats	2.0

Collection

Several collection methods are possible. Consider:
- Facility type.
- Labor requirements.
- Investment.
- Total manure handling system.

Slotted Floors

Concrete slats are the most durable and are suitable for animals of all ages. Both conventional reinforced and prestressed slats are available. Concrete slats can be purchased precast, precast at the building site, or cast-in-place. Some companies form gang slats to reduce installation labor.

Use slightly crowned and tapered slats (top width greater than bottom width) to improve cleaning. Provide a slip resistant surface for better footing and wear. See the concrete chapter. Edge homemade slats with a ¼" sidewalk edger to prevent chipping, improve cleaning, and reduce foot injuries. Allow for ¼"-½" clearance at each end of the slats for ease of installation. Provide a 1¾"-2" clear opening between slats adjacent to curbs to reduce solids buildup.

Use 6"-8" wide slats with cows and calves. Animals are usually dirtier on wider slats. Correct animal density can improve animal cleanliness. Provide a slot width of 1¾"-2" wide for cows and calves. A 1¼" slot with calves can reduce foot injury.

Scrapers and Blades

Either solid or semi-solid manure can be mechanically scraped. Remove manure on a regular schedule for cleaner buildings and livestock. Scraper systems create fewer odors because of frequent manure removal.

A mechanical scraper system has one or more scraping blades, a cable or chain to pull the scraper, and a power unit with controls. Common scraper systems are gutter cleaners and alley scrapers. A gutter cleaner has closely spaced flights on a chain drive in a narrow gutter (about 16") designed to handle manure with a high solids content. An alley scraper is a solid bar that pushes manure to the end of the barn. The cable or chain is usually recessed into the gutter to help stabilize the blade. In freestall barns, alley scrapers usually scrape the entire alley in one pass.

Mechanical scrapers are sometimes controlled with a time clock. Operate two to four times per day to keep manure from drying and adhering to the floor. More frequent or continual operation may be necessary during freezing conditions. In cold barns, install floor heat under the scraper rest position.

Mechanical scrapers can reduce daily labor requirements. However, maintenance requirements can be higher because of corrosion and deterioration due to the environment.

A small tractor with a back- or front-mounted blade or skid-steer can be used to scrape manure. They usually are more dependable and work better during cold weather.

Front-End Loaders

Front-end loaders remove solid manure from open lot surfaces, storages, and building floors.

Typical tractor loaders are available in 1,000 to 4,000 lb sizes. Because their turning radius is relatively large, limit their use to straight runs and areas with few turns. Wide front tractor wheel spacing is desirable for stability. A loaded bucket reduces rear wheel traction, so limit its use to relatively flat areas. Avoid building layouts requiring backing down long alleys or restricted turning areas.

Skid-steer loaders can clean in cramped areas, greatly reducing hand labor. Most have a turning radius of their own length and a low height, so they can work easily in tight quarters. Some have a relatively low load lifting capacity.

Flushing Systems

In a flushing system, a large volume of water flows down a sloped alley and carries manure to an outside storage. Wide open gutters can be used in free stall barns, holding areas, and milking parlors.

For free stalls, alleys are usually 8'-12' wide. Slope them 1%-4% to get initial flush water flow depths of 4"-6". Make the alley flat or form a ¼"-½" crown to force the flow towards the curbs. Make the curbs along the flush alley about 10" high. Elevate cross alleys to make alley curbs continuous. If the curbs are interrupted for cross alleys, flush water loses velocity and deposits manure.

Typically, 100 gal/day per 1,000 lb liveweight is needed for adequate cleaning. Flush with at least 75 gal/ft of alley width. Flush tanks should release the entire volume in 10 to 20 sec.

Flushing system performance can be a problem when outside temperatures stay below 20-25 F. Freezing usually occurs first around the tank and discharge channel rather than the alley. If the flush tank and discharge channel are both located inside the barn, flushing may be possible all winter in mild climates.

Usually flushed manure is stored in an earth storage basin and periodically pumped for irrigation. Often, solids settle out and are difficult to agitate and resuspend. Consider using a liquid-solid separator device before flushed manure goes to storage. Basin water can be reused for flushing when cows are not in the barn. With recycled water, salt and mineral concentrations can cause pumps and distribution pipes to plug. Check with your local milk inspector for approval.

Gravity Flow Channel

Gravity flow channels are rectangular-shaped channels with a flat, level bottom and a 6"-8" high dam at the discharge end, Fig 8-2. The dam retains a "lubricating" layer and the manure surface forms a 1%-3% incline. Manure flows by gravity down the inclined surface so no mechanical equipment is needed. Biological activity helps to liquify the manure and maintain a constant flow of manure over the dam into a cross channel or discharge pipe. In cold housing, these systems can freeze.

Channel width does not affect performance so it can be used under slotted floors. However, in stall milking barns limit channel width to 36"; 30" width is recommended.

Channel depth depends on channel length and manure surface incline. For design, assume the manure surface forms a 3% incline. Channel length typically ranges from 40'-80'. If the channel is longer than 80', consider incrementing the length into steps, Fig 8-3. See Table 8-4 for recommended step heights and channel depth. In stall milking barns, the maximum gutter depth at the end opposite the cross channel or discharge pipe must not exceed 54".

The overflow dam can be concrete blocks, a steel plate, or pressure preservative treated wood. Dam construction must be watertight. If dams are removable, it will allow for total cleanout.

Before putting cows in the building, fill the channel with 3"-6" of water to form the "lubricating" layer. Limit the use of bedding—up to 1 lb/cow per day of fine cut bedding. Do not use long straw bedding. Do not add milking center water containing disinfectants to the channel because it can kill the biological activity. Contact your milk inspector for approval before installation.

Gravity Transfer To Storage

Gravity flow transfer uses the hydraulic head exerted by relatively liquid waste to force wastes to flow. Pipe size to storage is determined by how liquid the

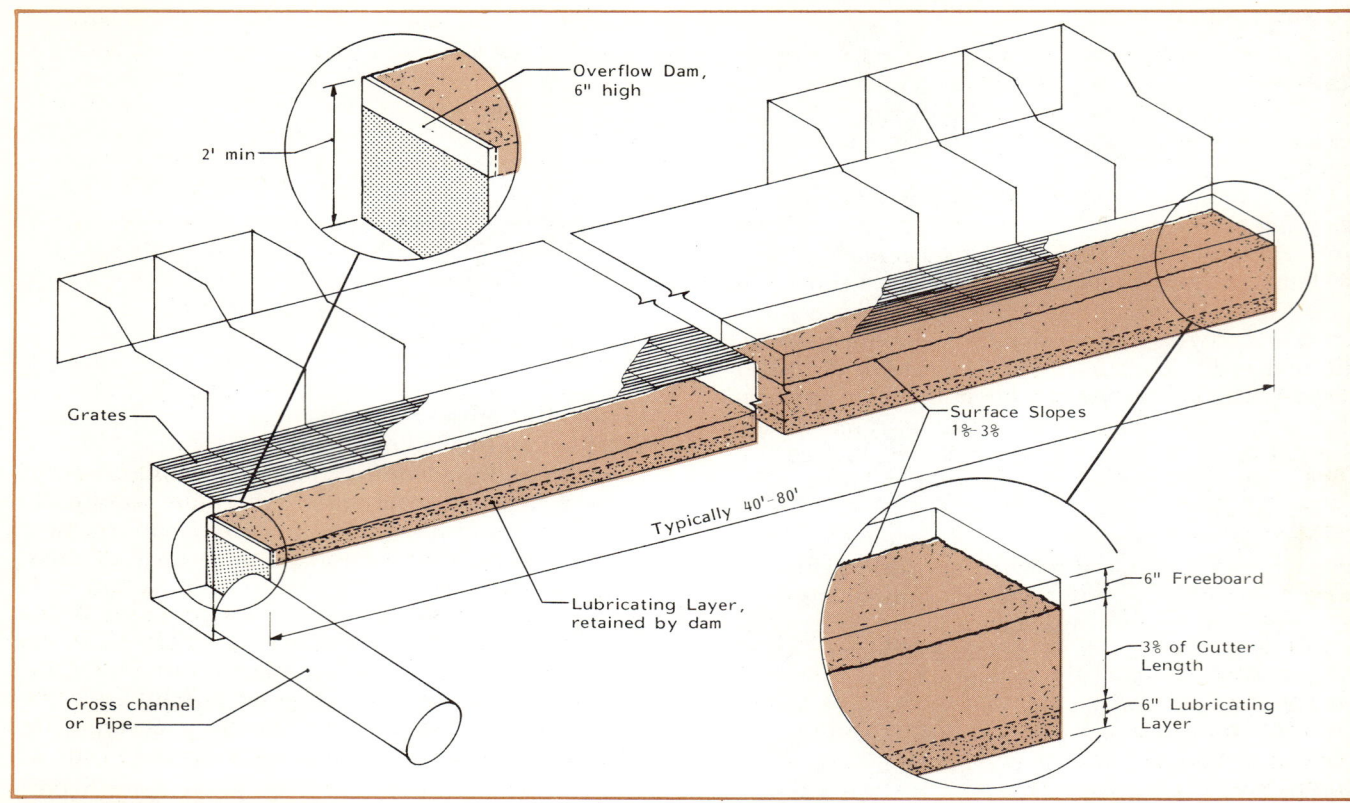

Fig 8-2. Gravity flow channel.
Make the cross channel at least 2' deep from the top of the dam to prevent manure from backing up the main channel.

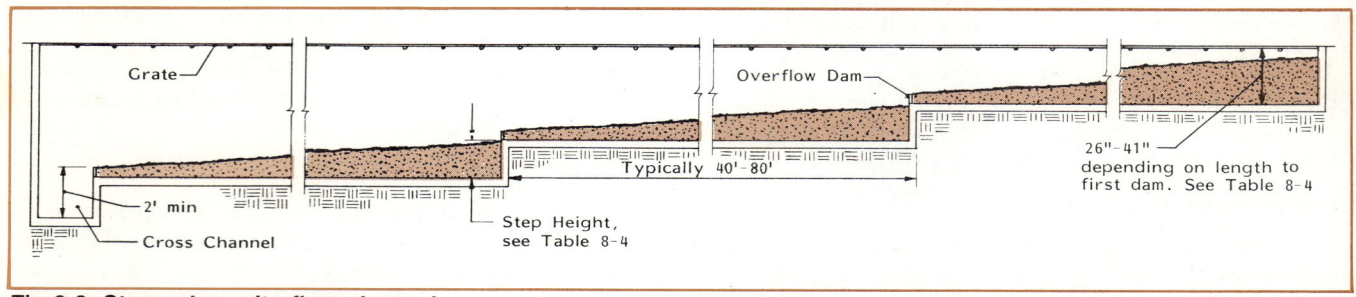

Fig 8-3. Stepped gravity flow channel.

Table 8-4. Gravity flow channel depth and step height.
Channel depth includes an allowance for a 6" dam and 6" freeboard.

Channel length ft	Channel depth in.	Step height in.
40	26	14
50	30	18
60	34	22
70	37	25
80	41	29

manure is. In general, two pipe size catagories are practical. Small diameter (6"-8") pipes work well with wastes which are quite liquid, such as milkhouse waste. Little or no bedding should be in the waste. Large diameter (20"-36") pipes transfer manure with well mixed bedding at a quantity up to about 3 lb/cow-day. Manure with excess bedding or long hay may not flow well unless water is added or elevation is increased.

Concrete, steel, or plastic pipe can be used to transfer manure. Slope small diameter pipes the entire length of the pipe. Level pipes lead to solids settling out and plugging. Slope 6" pipes a minimum of 1% and 8" pipes ½%. Avoid 90° bends in pipe; use wide sweep 45° elbows if turns are necessary. Install pipe below the frost line to prevent freeze-up. In northern climates, bottom load the storage, because top loading leads to freezing problems. In bottom loading systems, keep the end of the pipe 1'-2' off of the floor to reduce settled solids blocking the pipe. Cover the pipe end with at least 1' of liquid at the time of winter freeze-up. Provide cleanouts every 200' and at any bends.

Storage

Consider all farmstead operations, building locations, and prevailing summer winds when planning

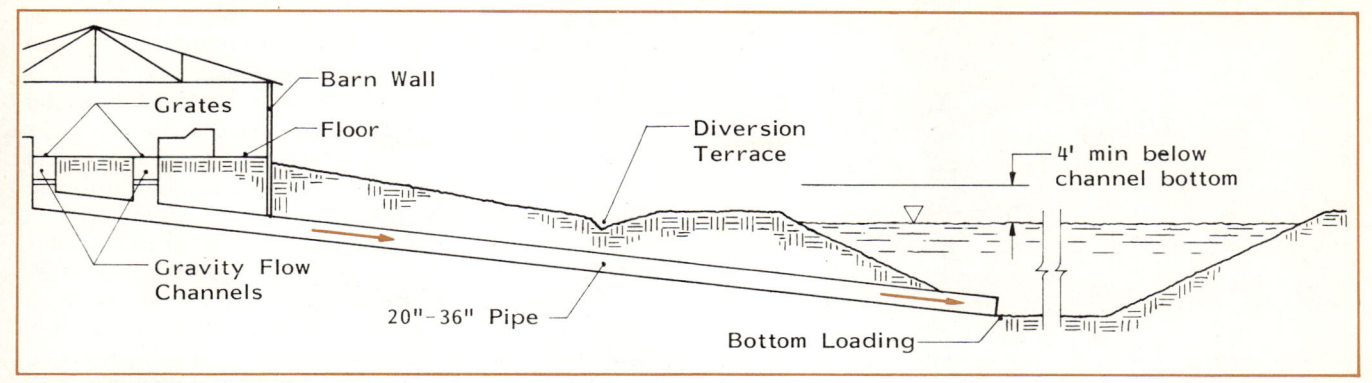

Fig 8-4. Manure transfer to storage.

storages. Allow at least 100' between a water supply and the nearest part of a storage. Locate manure storages at least 50' from a milkhouse or milking parlor. Check with milk and health authorities for minimum spacing requirements.

Evaluate site and soil conditions carefully to avoid contaminating ground and surface water. Avoid locating unlined storages over shallow creviced bedrock, below the water table, in gravel beds, or other areas where serious leakage can cause ground water pollution. Depending on local pollution control regulations, keep the bottom of the storage at least 3' above bedrock and at least 2' above the water table. Contact your county extension office or SCS for help in site evaluation.

Locate, size, and construct storages for convenient filling and emptying and to keep out surface runoff. Provide all-weather access and a firm base for the tanker loading area. Design for drive-through loading to avoid backing or maneuvering the tank wagon.

Liquid Manure

Storage capacity depends on regulations, number and size of animals, amount of dilution by spilled and cleaning water, amount of stored runoff, and desired length of time between emptying. Provide enough storage to spread manure when field, weather, and local regulations permit.

Plan for 10 to 12 mo storage capacity. See Table 8-2 for dairy manure production values. Provide extra capacity for dilution water, rain, snow, and milking center effluent. If the storage receives only animal manure, add dilution volumes of up to 10% for waterer wastage, rain, and snow. Add up to 60% dilution volume to the animal manure production if milking center effluent enters the storage. Provide at least 1' of freeboard.

From 20%-100% of the storage volume may be needed for dilution water if the manure is to be irrigated onto cropland.

Below-Ground Storage

Liquid manure can be stored in below-ground tanks. Storage depth may be limited by soil mantle depth over bedrock, water table elevation, and possibly, effective lift of a pump.

Tanks must be designed to withstand all anticipated earth, hydrostatic, and live loads, plus uplift if a high water table exists. Columns and beams to support a floor or roof are usually spaced 8'-12' apart. For information about concrete tank design order TR-3, *Concrete Manure Tank Design* TR-4, *Welded Wire Fabric in Concrete Manure Tanks* TR-9, *Circular Concrete Manure Tanks* and plan mwps-74303, Liquid Manure Tanks — Rectangular, Below Grade from the addresses listed inside the front cover of this book. SCS offices can also provide concrete tank designs.

In cold climates, insulate the upper 2'-3' of exterior tank walls with waterproof insulation. When construction is completed, clean out foreign material that could damage pumps. Before filling with manure, add 6"-12" of water to keep manure solids submerged and counteract the uplifting forces caused by external pressures.

Earth Storage Basins

Earth basins are earth-walled structures at or below grade which may or may not be lined. They can provide long-term storage at a low to moderate investment. They are designed and constructed to prevent ground and surface water contamination. Check with your SCS office for help in evaluating site suitability, dike construction, bottom sealing, and basin wall sideslopes. In general, steeper banks conserve space, reduce the amount of rainfall runoff entering the storage, and leave less manure on the banks when emptied. Inside bank slopes of 1:2 to 1:3 (rise:run) are common for most soils. Make the outside sideslopes no steeper than 1:3 for easier maintenance. Make the embankment wide enough (at least 12') for access by mowers and agitation equipment.

Above-Ground Storage

Above-ground circular manure tanks are usually short, large-diameter tanks resembling silos. They are expensive compared to earth basins and are usually not used to store runoff or dilute wastes. However, they are a good alternative where basins are limited by space, high groundwater, shallow creviced bedrock, or where earth basins are not aesthetically acceptable.

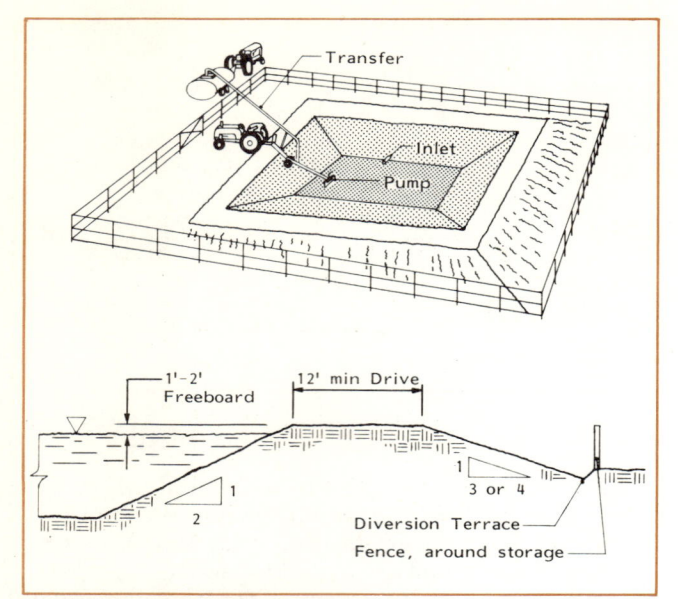

Fig 8-5. Earth storage basins.

Above-ground liquid storages are from 10'-60' high and 30'-120' in diameter. They are made of concrete stave, reinforced concrete, and steel. Leaks from joints, seams, or bolt holes can be unsightly, but most small leaks quickly seal with manure. The joint between the foundation and the sidewall can be a problem with improper construction. The reliability of the dealer and construction crew are as important as the tank material in assuring satisfaction.

Filling

Manure can be transfered to storage by pumping or gravity. With outside storages, locate collection pits and sumps where manure can be conveniently scraped in. Water may need to be added for easier transfer.

Use bottom loading for wastes that form a surface crust. The crust reduces fly and odor problems. Keep the inlet pipe about 1' above the bottom to prevent blockage, freezing, and gases from leaking back into the barn.

Top loading is suitable for a storage that will not crust over. Top-loaded solids pile up around the loading point, because the manure does not flow away, particularly in cold weather. Bottom loading pushes solids away from the inlet and distributes them more evenly.

Agitation

Agitate liquid manure storages by diverting part or all of the pumped liquid through an agitator nozzle. The liquid stream breaks up surface crust, stirs settled solids, and makes a more uniform mixture. Agitiate and pump as much manure as possible, then agitate the remaining solids and dilute if necessary. Diluting wastes to 90% water may be necessary. Keep surface runoff out of the storage. Use roof runoff for dilution.

Agitation pumps include open and semi-open impeller centrifugal chopper agitator pumps and helical screw pumps. For more information about pumps, see the section on pumping manure.

Earth basins are usually agitated with a three-point hitch or trailer-mounted high capacity manure pump mounted at the end of a long boom. The pump is lowered down the embankment or ramp, placing the pump inlet under the manure surface. These pumps have a chopper and/or rotating auger section at the inlet to break up the crust and draw solids into the pump for thorough agitation

Above-ground storage and manure tanks are usually agitated with a centrifugal pump. With silo storages, pumps can be mounted on the storage foundation. Large diameter tanks may have a center agitation nozzle. Agitate below-ground manure tanks with a submerged centrifugal pump, Fig 8-6. Locate agitation sites no more than 40' apart in tanks without partitions (20' o.c. for vacuum pumps). With below-building storage, a continuous 6' wide annex the length of the building can be used, Fig 8-7. Locate support columns so they do not interfere with agitation.

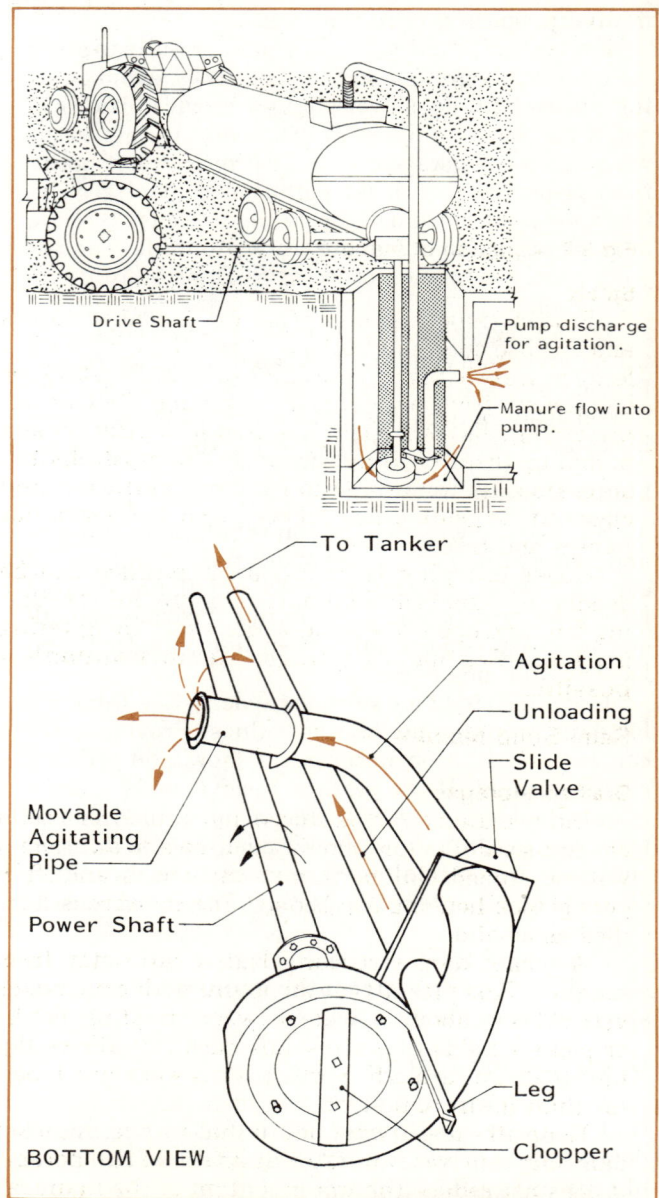

Fig 8-6. Agitating with submerged centrifugal pump.

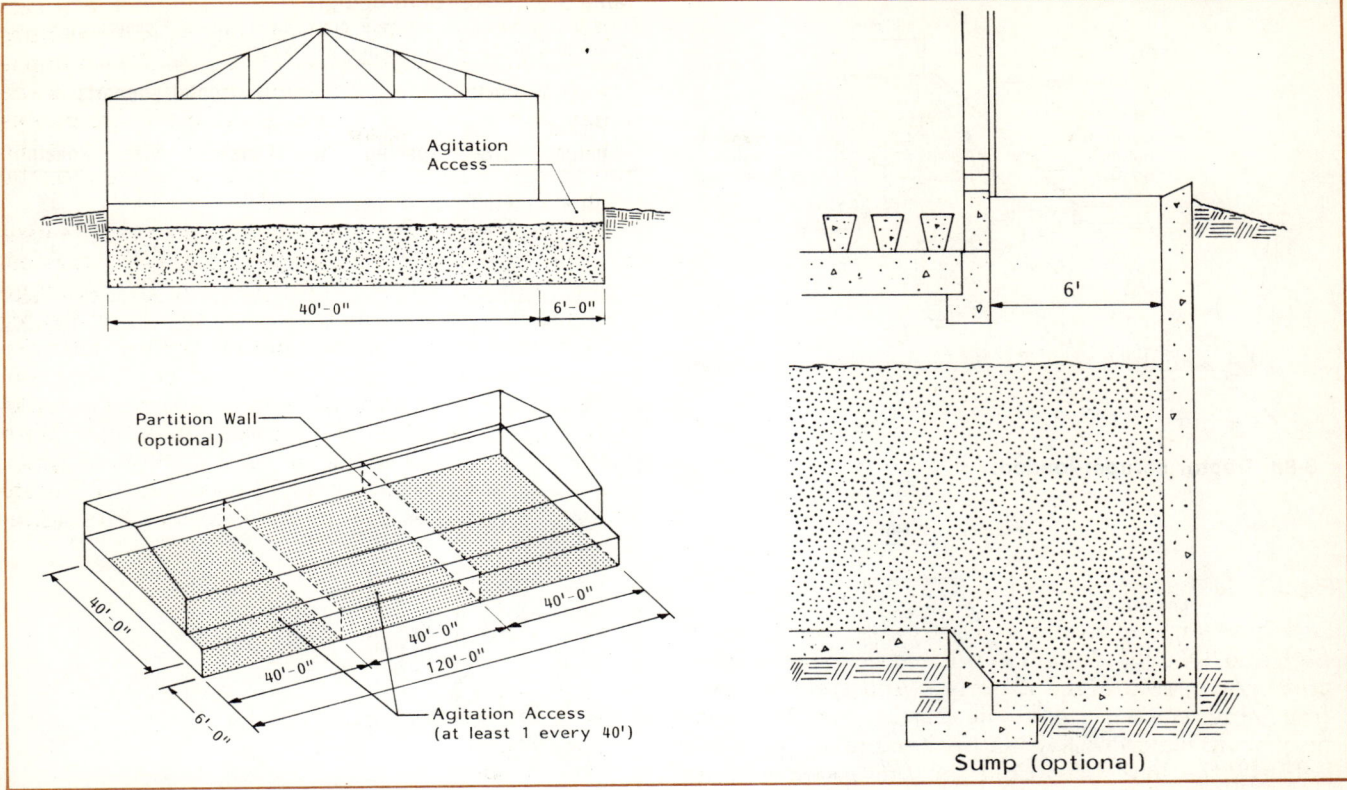

Fig 8-7. Agitation of manure storage pits.

Safety

Protect tank openings with grills and/or covers and enclose open-top tanks and earth basins with a fence at least 5' high to prevent humans, livestock, or equipment from accidentally entering. Provide removable grills over openings used for agitation and pumping. Install railings around all pump docks and access points for protection during agitation and cleanout. Provide wheel chocks and tie downs for pumps and tractors.

Gases escaping from agitated manure can be deadly to animals and humans. Operate all ventilating fans and open doors and windows when agitating and unloading manure storages. **Remove animals if possible.**

Semi-Solid Manure

Drained storages

Dairy manure can be stored and hauled as a solid or semi-solid if ample bedding is used and additional water excluded. Calf pens, maternity pens, and often youngstock housing are bedded and manure is handled as a solid.

A picket dam structure drains rainwater from manure. This picket-type structure with continuous vertical slots about ¾" wide between standing planks or pickets holds manure solids back but allows liquids to drain through. Vertical slots work much better than horizontal slots.

Keep all excess water out of the storage. A picket dam only removes rainwater that falls on the storage; it does not reduce the water content of the manure.

These manure handling systems require that bedding such as straw or sawdust be added to the manure to provide for convenient handling with conventional spreaders. Do not use this storage for sand-bedded free stalls.

The drainage water from the manure storage is polluted; keep it from entering public waters, polluting groundwater, or leaving your property. Direct drainage water to a holding pond or earth basin.

Roofed storage

Roofed storages were developed for dairy stall barns where large amounts of bedding are mixed with manure. This system provides an aesthetically pleasing structure which appears to be another building on the farmstead.

Above-ground roofed storages have a concrete floor as much as 3' below existing grade. They usually have 12' post-and-plank or concrete walls with earth backfill to withstand internal fluid pressures, Fig 8-9. The roof protects collected manure from rainwater to keep it as dry as possible. Manure in storage crusts over, reducing odor and fly breeding problems. Provide a ventilating space between the top of the walls and the trussed roof.

These storages are typically bottom-loaded with a large diameter piston pump.

Unload by lifting horizontal planks out from a door in one end of the storage. Manure then flows over the wall onto a concrete slab outside the storage where a front-end loader fills a conventional spreader. Remove the planks with a hoist supported from a beam above the doorway.

8-8a. Tractor scraper-loaded.

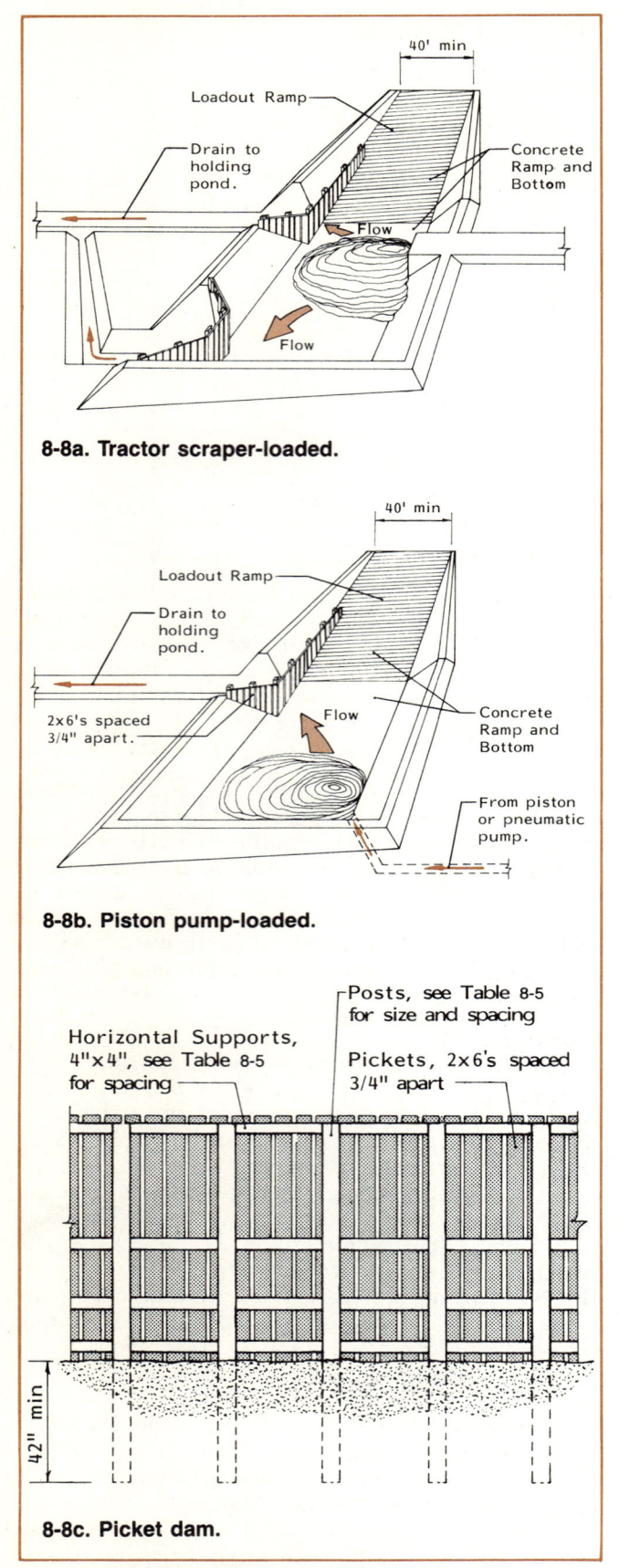

8-8b. Piston pump-loaded.

8-8c. Picket dam.

Fig 8-8. Picket-drained storage.

Table 8-5. Picket dam design.
Posts and horizontal supports are rough sawed timbers. Pickets are pressure preservative treated 2x6s; rough sawn or surfaced.

Picket height	Posts Size	Spacing	Distance from top of pickets	Horizontal supports Size	Spacing
0'-4'	4"x6"	5'	0'-4'	4"x4"	3'
5'	6"x6"	4'	4'-6'	4"x4"	2.5'
6'	6"x8"	4'	6'-8'	4"x4"	2'
7'	8"x8"	3'			

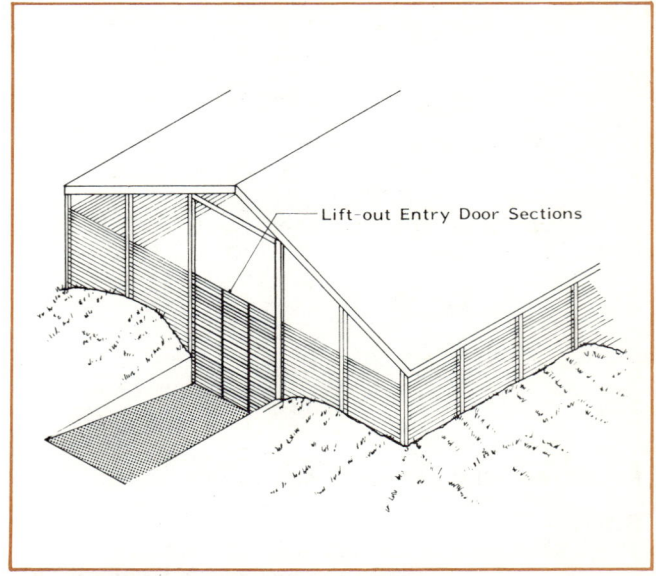

Fig 8-9. Roofed above-ground storage.

Solid Manure

Solid manure storage is a part of some manure handling systems. Provide for convenient filling with a tractor-mounted manure loader or scraper, elevator stacker, blower stacker, or piston pumping system. Unload with a tractor-mounted bucket. Locate storage for year-round access so manure can be spread when field, weather, and regulations permit. Prevent surface runoff water from entering the storage.

Adequate storage is needed for convenience, maintaining maximum fertility, and preventing water pollution. Storage capacity depends on number and size of animals, type and amount of bedding, and desired storage period. Up to about 180 days of storage is recommended in cold climates, with less needed in warmer climates.

Provide convenient access for unloading and hauling equipment. Slope entrance ramps upward to keep out surface water. Provide a load-out ramp at least 40' wide with a 1:10 slope (1:20 preferred). A roughened ramp improves traction. Angle grooves across the ramp to drain rainwater. Concrete floors and ramps 4" thick are recommended. Slope the floor ¼"/ft (2%).

Walls are usually concrete or post-and-plank. Provide one or two sturdy walls to buck against for unloading.

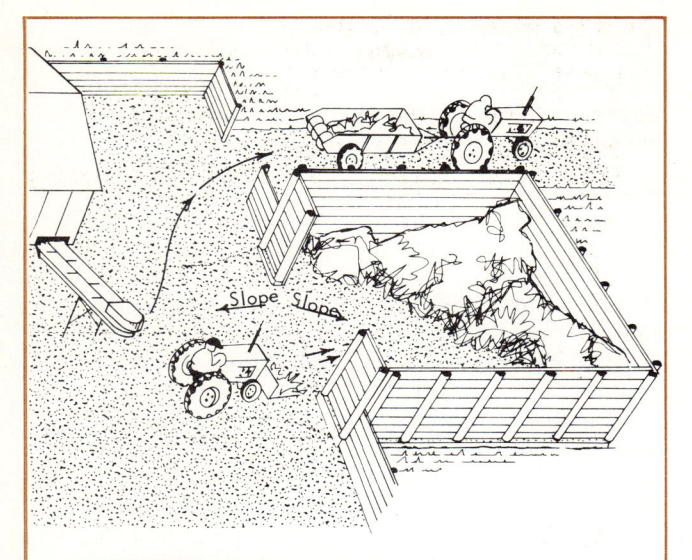

Handling Manure

Solid Manure

Most solid manure spreaders are box-type. Others include flail-type spreaders, dump trucks, earth movers, or wagons. A spreader should distribute manure uniformly. Spreader mechanisms include paddles, flails, and augers. The feed apron, which moves the waste to the spreader, is often variable-speed. Some spreaders have moving front-end gates that push the wastes to the spreading mechanism.

Flail-type spreaders usually have a shaft mounted near the open top and parallel to the main axis of the tank. Chain flails on this shaft throw the manure out the side of the spreader as the shaft turns.

Front-end loaders, scrapers and blades, and several mechanical systems transport solid manure, but are not usually used for spreading.

Liquid Manure

Manure with up to about 4% solids can be handled as a liquid with irrigation or flushing equipment. From 4%-15% solids, manure can be handled as a liquid, but equipment needs differ. Fibrous materials, such as bedding, hair, or feed, can hinder manure pumping. Chopper pumps can cut fibrous materials for improved pumping. Piston manure pumps can handle manure with bedding.

Liquids—up to 4% solids

Liquid manure after solids have been separated or manure with dilution water added can have 4% or less solids. With proper management and screening, a liquid pump is satisfactory, but a slurry or trash pump is more trouble-free. If large quantities are handled, a pipeline may be preferred over tank wagons for transport.

Settle out solids if possible. For irrigation, provide an intake screen on the pump, with screen openings no larger than the smallest sprinkler nozzle, spile tube, or gate. A large screen area reduces plugging and velocities into the screen.

Protect pumps and power units against malfunctions such as plugged lines or nozzles, loss of prime, overheating, and loss of lubricant. Match safety sensors and controls to pump, motor, and system types. Safety devices include temperature and pressure gauges, fuses, and circuit breakers.

Increase the size of power units rated for water capacity by at least 10% for increased friction losses and specific gravity of livestock wastes. Select pumps adequate for the flow and head of your system.

Slurries—4% to 15% solids

Solids and liquids separate, so agitate manure before pumping. Use open impeller chopper pumps to agitate slurries in storage.

Pumps moving slurries through long pipelines and sprinklers might operate against fairly high pressures. Pumps for furrow irrigating with gated pipe or for filling tanks usually operate against much lower pressures.

8-10a. Spreader or tractor stacking.

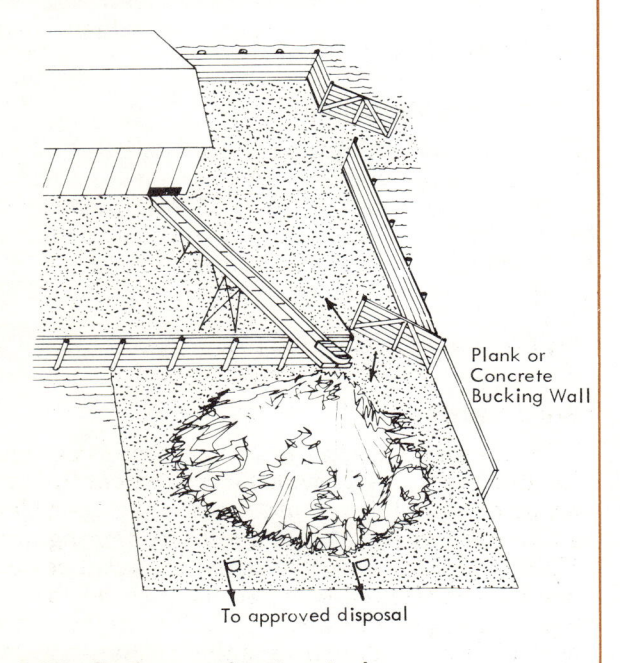

8-10b. Stationary elevator stacker.

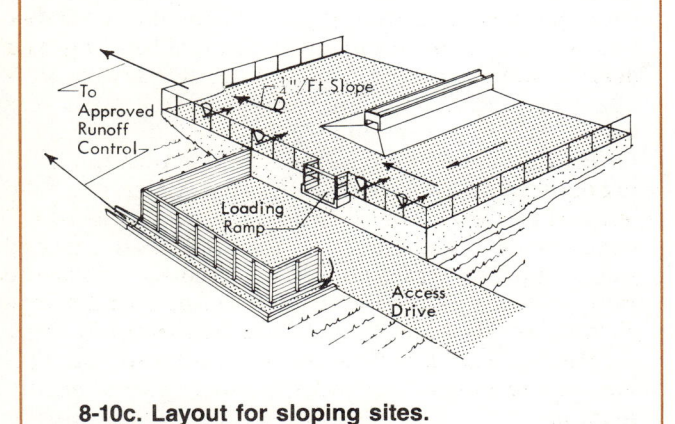

8-10c. Layout for sloping sites.

Fig 8-10. Solid manure storages.

Piston, helical rotor, submerged centrifugal, and positive displacement gear-type pumps can handle heavy slurries against high pressures. However, their performance is improved if solids are below 10%. They do not require priming and therefore adapt to automation. To pump heavy slurries against low pressures, use submerged centrifugal, piston, auger, or diaphragm pumps.

Irrigation

Irrigation equipment distributes water and fertilizer on crops. Most irrigation systems can handle liquid manure with up to 4% solids, which is typical of lot runoff and effluent from a lagoon, holding pond, or milking center. Surface irrigation and big guns can handle higher solids content fluids.

Surface spreading is an effective method of manure disposal. It has low cost, low power requirements, and few mechanical parts. Do not use surface irrigation on land with more than a 2% slope.

Sprinklers allow waste disposal on rolling and irregular land. Although initial and operating costs are generally higher for sprinklers than for surface systems, labor requirements are reduced, systems can be automated, and application uniformity is improved.

Select an irrigation system adapted to your topography, soil, and crops. A well designed and managed system prevents runoff and erosion.

Once you decide to irrigate for disposal and have an idea of the system you want, get MWPS-18, *Livestock Waste Facilities Handbook,* for more information about system selection, design, and management. Additional help can be obtained from the Extension Service, SCS, consulting engineers, or irrigation equipment dealers.

Handling Lot Runoff

Collecting Runoff

Lot runoff, whether from rainfall or snowmelt, contains manure, soil, chemicals, and debris and must be handled as part of the manure management system.

Runoff from roofs, drives (but not animal alleys), and grassed or cropped areas without livestock manure is relatively clean and need not be handled in the manure management system. Divert clean runoff either away from the system to reduce the total volume or into the system for dilution. Use curbs, dikes, culvert pipes, and terraces to divert clean water away from a manure area. For help with runoff control, consult your local SCS office.

Liquid-Solid Separation

Liquid-solid separation is accomplished by gravity, centrifuges, particle-size screens or filters, or evaporation of water. After solids are removed, they can be spread as a fertilizer and soil conditioner, used for bedding, or recycled as livestock feed.

Mechanical separators require at least a pump to

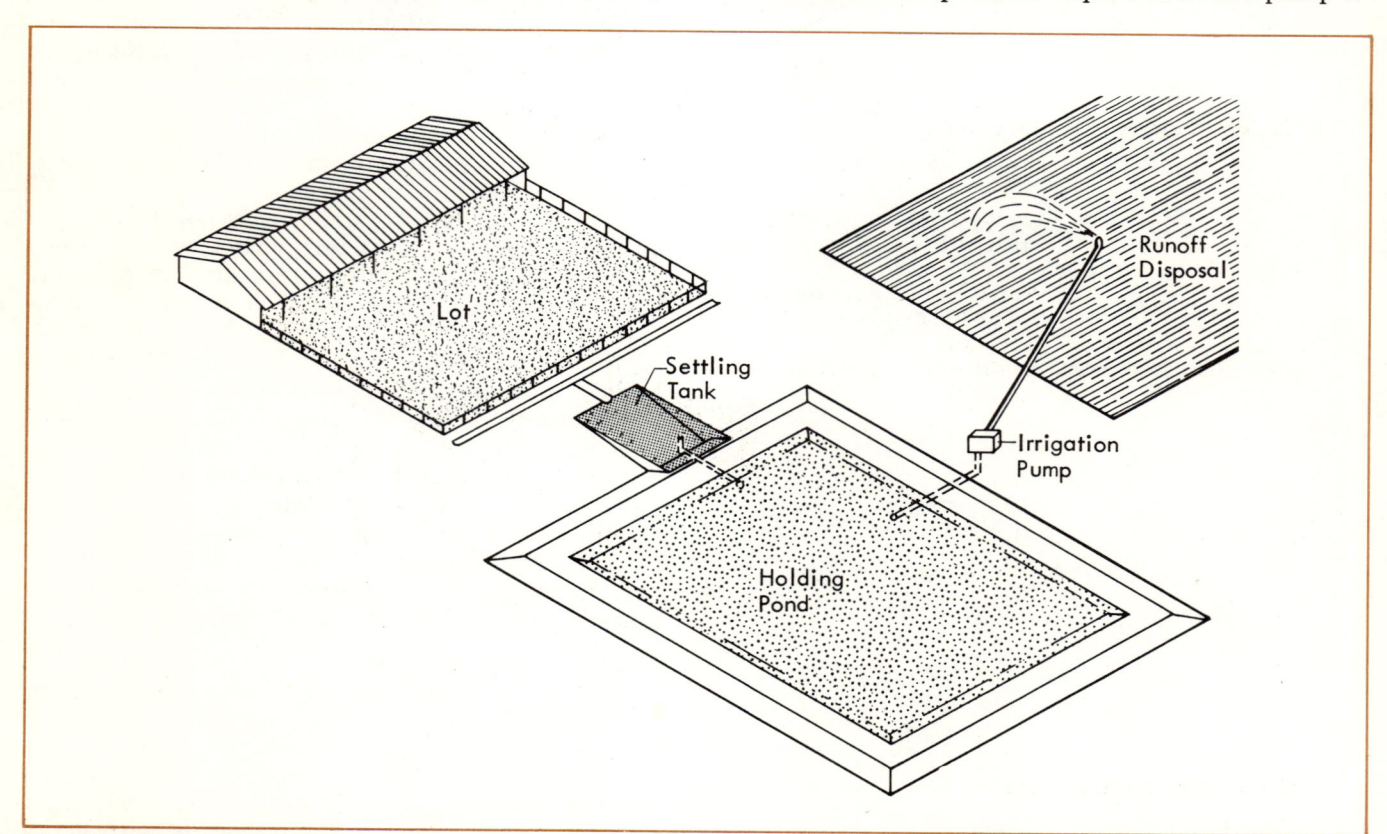

Fig 8-11. Lot runoff handling system.

deliver the manure slurry to the separator. Separator units use screens and rollers to separate solids from liquids. The power requirements of a separator can vary up to 25 hp depending on the type. Several commercial mechanical liquid-solid separators are available. Consult your state extension agricultural engineer for assistance.

Settling basins are an effective gravity liquid-solid separator. Settling basins remove 50%-85% of the solids from lot runoff. Baffles and porous dams slow the flow enough so that settling occurs. Make the settling basin floor concrete to improve solids removal. Remove solids as needed between runoff events. For more design information, see MWPS-18, *Livestock Waste Facilities Handbook*.

Holding Pond

A holding pond temporarily stores runoff water from a settling basin, Fig 8-13. A holding pond does not receive roof water, cropland drainage, or "clean" water from other sources. Design for at least the

8-12a. Earth sidewall settling basin.

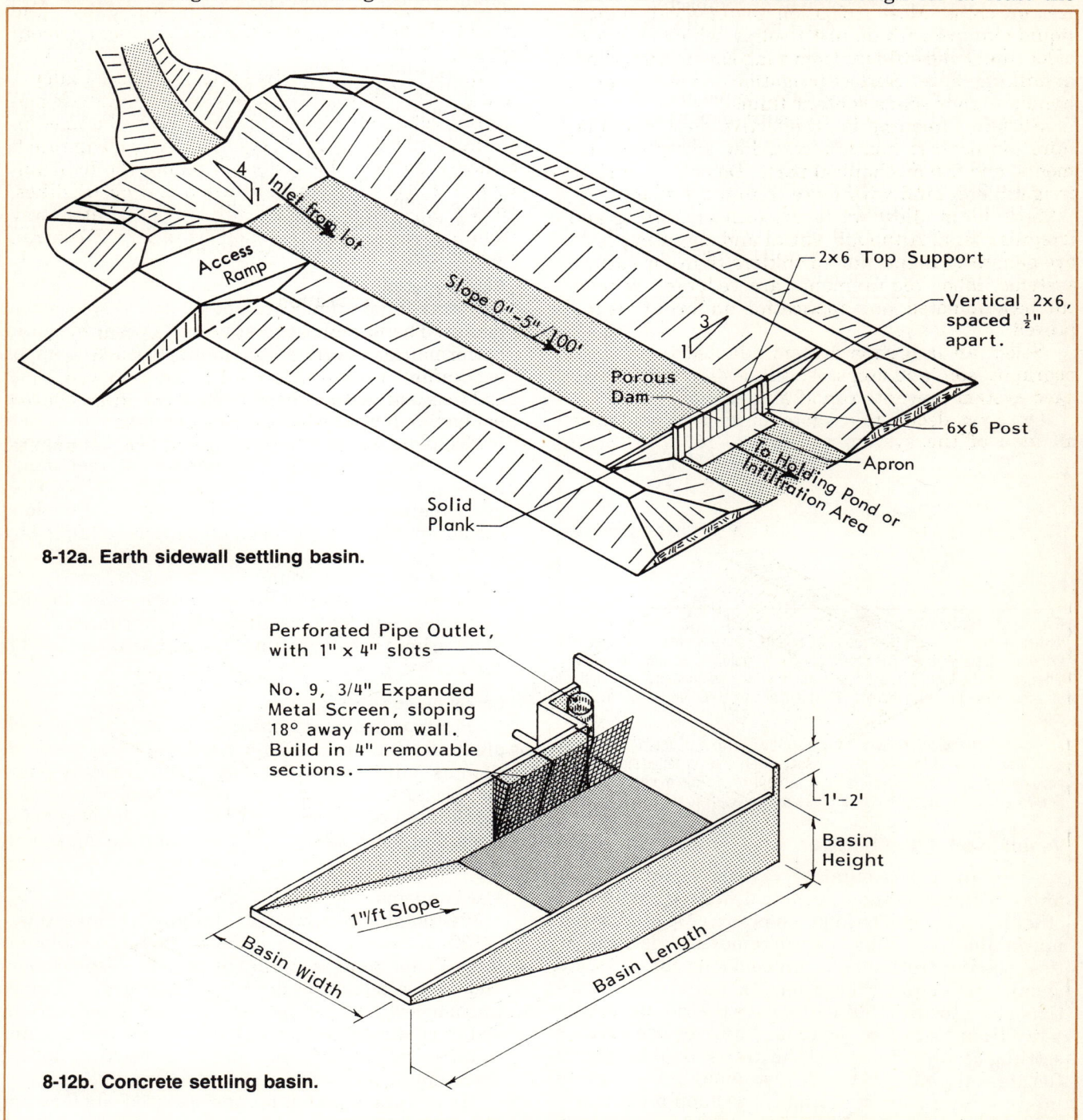

8-12b. Concrete settling basin.

Fig 8-12. Settling basins.

storage time required in your state (usually 90 to 180 days). Base capacity on inches of rainfall on a drained area, such as a 25-yr, 24-hr rainfall event. With additional capacity, emptying the pond can be delayed to fit labor and cropping schedules without fear that another runoff event will cause overflow.

Features of a holding pond:
- The bottom and sides tend to seal naturally. If the pond is in sandy or gravelly soils or near fractured bedrock, seal the bottom by lining with an approved plastic or 6" of compacted clay.
- It receives runoff from a lot, usually after it has been through a settling unit, or from a lagoon's overflow.
- The interior banks of earth dams are usually no steeper than 1:2.5 (rise:run), depending on soil type. To maintain sod and mow weeds, limit exterior slopes to no steeper than 1:3.
- Empty the pond by pumping, usually through irrigation equipment.
- Pump before the storage is full and when liquids will infiltrate into the soil. Avoid spreading on frozen or wet ground.

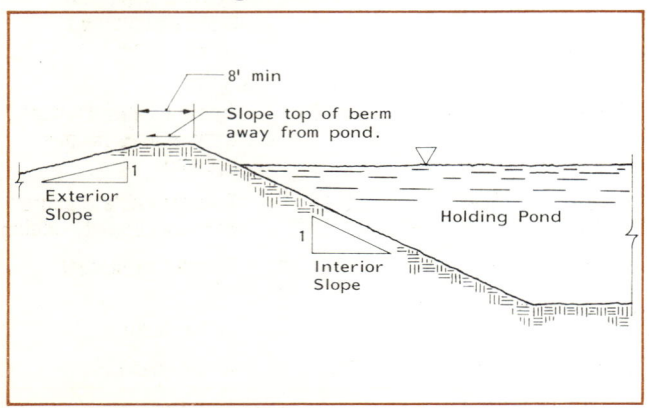

Fig 8-13. Holding pond.
Vertical dimensions are exaggerated to emphasize slopes. The steepest exterior sideslope is 1:3 (rise:run); 1:4 if it is going to be mowed. Interior sideslope varies from 1:1 to 1:4 depending on the soil construction characteristics. Make the berm at least 12' wide if agitation equipment will travel on top.

Provide an emergency spillway in undisturbed earth near the holding pond dam. Make it 1'-2' below the top of the dam. Overflow from liquid storages is a potential pollutant, and therefore must be controlled. Run the spillway overflow to secondary treatment or a grassed disposal field.

Vegetative Infiltration Area

Lot runoff treatment by vegetative infiltration areas is practical and economical when grassland or other land that can be kept out of row crop production is available. A settling basin to remove solids is essential. The infiltration area can be a long, 10'-20' wide channel or a broad, flat area. For uniform distribution, slope the first 50' 1% to move the runoff rapidly away from the lot and settling unit before further settling occurs. The rest of the area should be nearly flat (less than 0.25%) with just enough drainage to prevent water from standing. The runoff must travel through the vegetated area at least 300' before entering a stream or ditch.

For more detailed information, see MWPS-18, *Livestock Waste Facilities Handbook,* or contact your local SCS office.

Pumping Manure

Pump Selection

Solids content and required pumping pressure are the major factors in selecting a manure handling pump, Table 8-6. Sand is hard on pumps. Do not use pumps with a sand bedded system. Manure pumps often require repair. Correctly size the pump to reduce breakdowns and need for repair.

Solids content varies with animal age, ration, housing type, and collection system. Settle out solids or add dilution water to reduce pumping problems.

Pressure requirements vary considerably with the intended use and application. A sprinkler irrigation system requires high pumping pressures, but other systems may only need to lift manure 10'-20' to a manure tanker or earth storage.

Centrifugal pumps

Centrifugal pumps are not positive displacement pumps. The impeller can slip in the pumped liquid. Performance of a centrifugal pump depends on impeller design.

Closed impellers are efficient with water but cannot handle large solid particles. Semi-open impellers have a plate or shroud on only one side of the impellers and handle liquids with some solids. Open impeller pumps handle liquids with up to 15% solids. They are sometimes fitted with a sharp rotating blade at the pump inlet to chop material such as bedding, hay, or silage.

Most centrifugal pumps for livestock waste handling are self-priming. But with high-solids liquids, foot valves can leak, requiring hand priming. Avoid hand priming by installing pumps below the level of the pumping reservoir.

Screw pumps

Positive displacement screw or rotary pumps handle fluids that are free from hard and abrasive solids such as nails and stones. The common rotary pump for livestock manure is the helical screw.

Do not operate these pumps dry. Direct a small stream of fresh water into the pump casing during operation.

Piston pumps

Piston pumps can handle liquids high in solids—up to whatever can flow or be pushed into the pump inlet. Piston pumps operate by positive displacement and have a pulsating discharge. Manure is scraped into a below grade hopper and flows by gravity to the piston chamber. One-way check valves at the piston chamber exit prevent manure from backing up into the pump.

Two types of piston pumps are used—hollow and solid. Hollow pistons handle fairly dilute manure while the more expensive solid piston pumps handle

Table 8-6. Manure handling pumps.

Pump type	Maximum solids content %	Agitation ability	Agitation range ft	Pumping rate gpm	Pumping head ft of water	Power requirements hp	Applications
Centrifugal							
Open & semi-open impeller							
Vertical shaft chopper	10-12	Excellent	50-75	1,000-3,000	25-75	65+	Gravity irrigation Tanker filling Pit agitation Transfer to storage
Inclined shaft chopper	10-12	Excellent	75-100	3,000-5,000	30-35	60+	Earth storage agitation Gravity irrigation Tanker filling
Submersible transfer pump	10-12	Fair	25-50	200-1,000	10-30	3-10	Agitation Transfer to storage
Closed impeller	4-6	Fair	50-75	500+	200+	50+	Recirculation Sprinkler irrigation
Elevator	6-8	None	0	500-1,000	10-15	5+	Transfer to storage
Helical screw	4-6	Fair	30-40	200-300	200+	40+	Agitation Sprinkler irrigation Transfer to storage Holding pond and lagoon pumping Tanker filling
Piston							
Hollow	18-20	None	0	100-150	30-40	5-10	Transfer cattle manure without long fibrous bedding
Solid	18-20	None	0	100-150	30-40	5-10	Transfer cattle manure with unchopped bedding
Pneumatic	12-15	None	0	100-150	30-40	—	Transfer to storage
Self loading tanker							
Centrifugal, open impeller	6-8	None	0	200-300	N/A	75+	Tanker loading
Vacuum pump	8-10	Poor	20-25	200-300	N/A	50+	Tanker loading

manure with large amounts of bedding, Fig 8-14.

Hollow pistons are rectangular and run in a loose-fitting square cylinder at the base of a sloping hopper, Fig 8-14a. Discharge pipes are usually 9"-15" wide.

Solid piston pumps have round, square, or triangular pistons. The piston is in line with the transport pipe and is located at the base of the hopper, Fig 8-14b. The close fitting piston creates a vacuum in the cylinder pulling manure into the cylinder.

Air driven pumps

Vacuum tank wagons and air displacement transport pumps are two types of air driven pumps for handling manure.

Vacuum tank wagons are filled by suction developed by a PTO driven vacuum pump. There are simpler spreader tanks filled by a separate pump. Vacuum tanks can pump liquid manure with up to about 10% solids.

Air displacement transport pumps have a large (about 1,500 gal) underground steel holding tank and a compressed air supply, Fig 8-15. Manure from a gutter cleaner or tractor scraper collects in the tank. To empty the holding tank, the loading hatch is closed manually or with an air cylinder and compressed air is pumped into the top of the holding tank. As pressure in the tank builds up, manure is forced out of the tank through a large underground PVC pipe to storage. The tank outlet has a flap check valve to prevent backflow.

Agitation pumps

Agitation pump types include open and semi-open impeller centrifugal chopper-agitators and helical screws. They can pump liquid manure with fibrous solids up to about 15%.

These pumps agitate by diverting part or all of the pumped liquid back through an agitator nozzle. The liquid stream from this nozzle breaks up surface crust, stirs up settled solids, and makes a more uniform slurry. Most chopper-agitator pumps can effectively agitate manure for a distance of about 40' from the pump.

Irrigation pumps for slurry

For irrigation slurries, semi-open impeller centrifugal and screw type pumps are common. Many of these pumps handle slurries with more than 10%

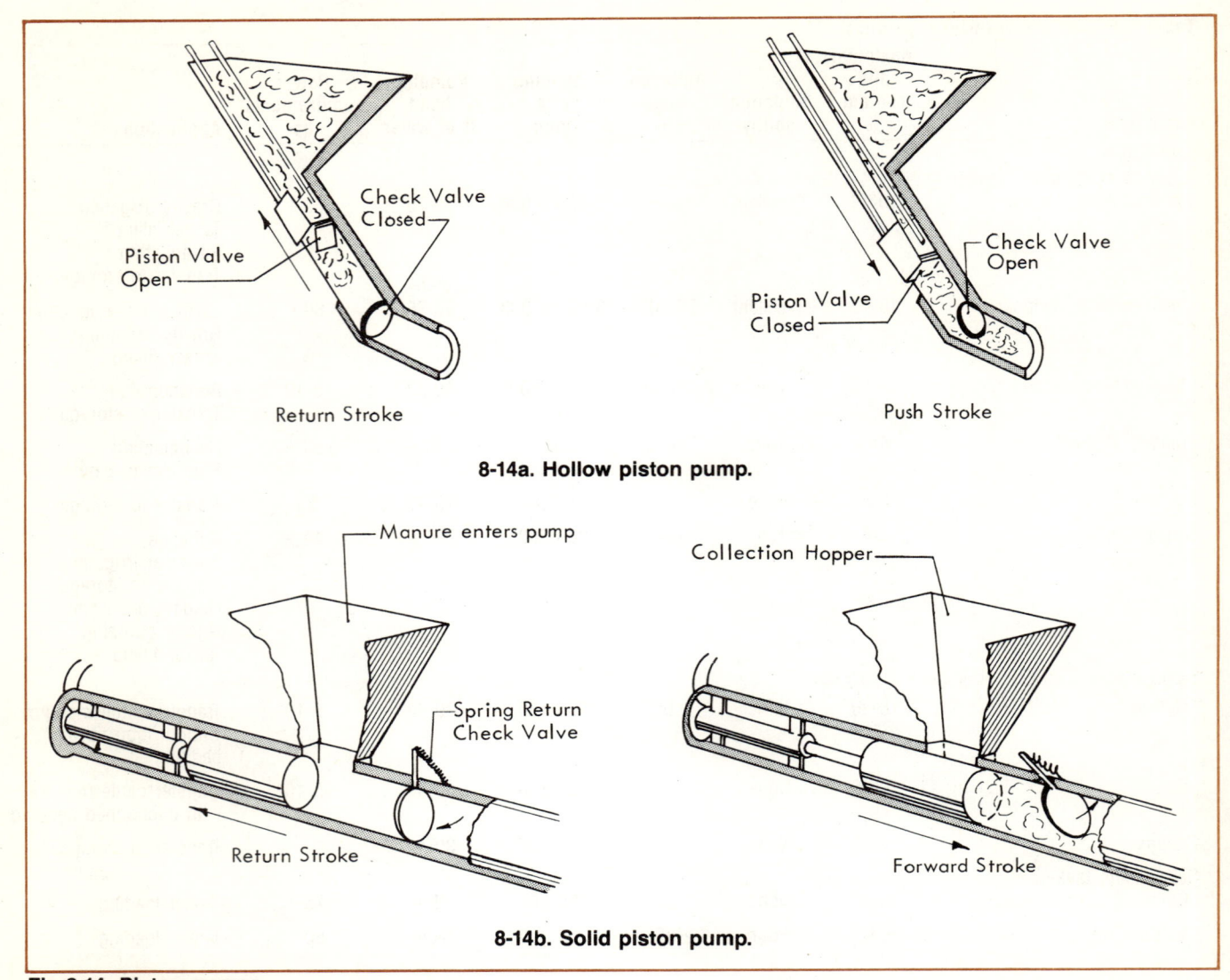

Fig 8-14. Piston pumps.

8-14a. Hollow piston pump.

8-14b. Solid piston pump.

solids, but the high pipeline friction losses usually make pumping these high-solids slurries impractical. Irrigation pumps work against fairly high pressures, especially with sprinkler irrigation. Sprinklers are usually the big gun-type with 7/8″ and larger nozzles, requiring pressures of 80 psi or more, and pumping rates of 100 to 800 gpm.

Irrigation pumps for low solids wastes

Effluents from lagoons, holding ponds, and milking centers usually have 3% solids or less. Single stage, standard closed impeller centrifugal pumps work well. Semi-open impeller pumps are sometimes desirable to reduce clogging. Screen out large solids at the pump inlet.

Flushing and recirculating pumps

Centrifugal pumps with semi-enclosed or enclosed impellers and positive displacement rubber impeller or roller pumps work well to recirculate lagoon water for flushing and diluting pit storages. Rubber impeller and roller pumps are capable of higher pressures than small centrifugal pumps.

A major disadvantage of recirculating pumps is the frequent service required when operated continuously. A plastic-fitted pump casing with high quality seals that can be easily serviced is recommended. Because of their short service life, stock several spare replacement seals, impellers, and a replacement motor.

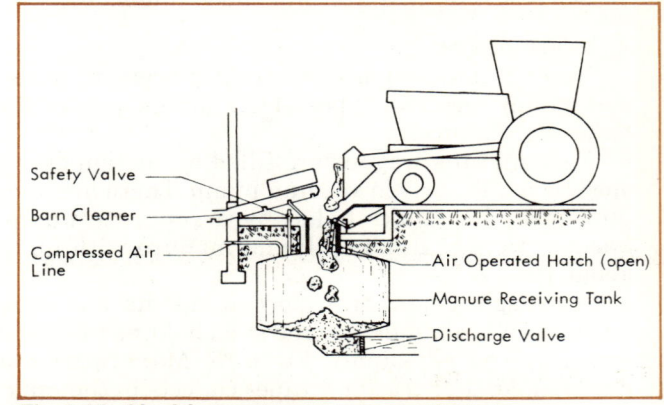

Fig 8-15. Air driven manure transport pump.

12.19

Stock Guards

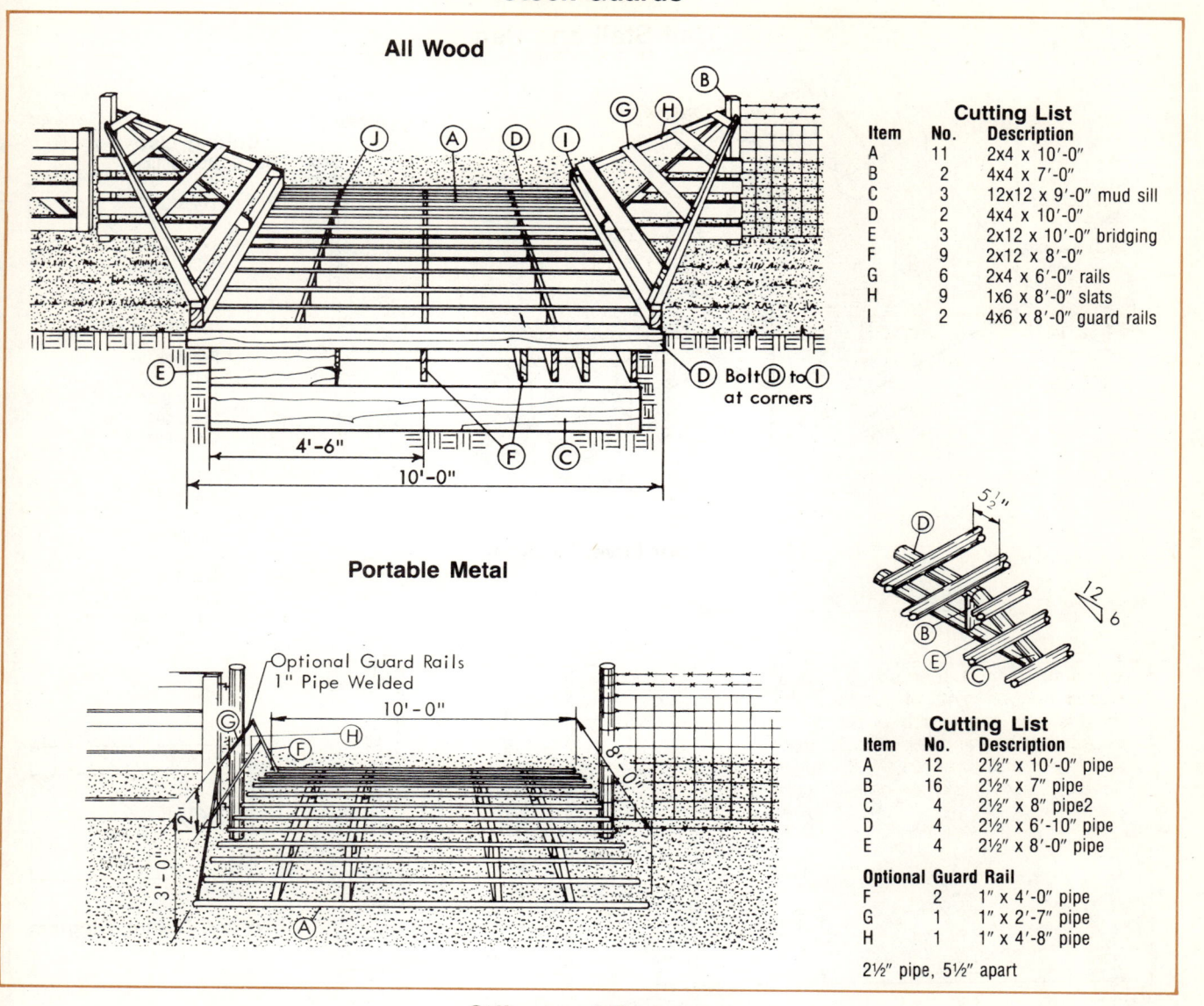

All Wood

Cutting List

Item	No.	Description
A	11	2x4 x 10'-0"
B	2	4x4 x 7'-0"
C	3	12x12 x 9'-0" mud sill
D	2	4x4 x 10'-0"
E	3	2x12 x 10'-0" bridging
F	9	2x12 x 8'-0"
G	6	2x4 x 6'-0" rails
H	9	1x6 x 8'-0" slats
I	2	4x6 x 8'-0" guard rails

Bolt D to I at corners

4'-6"
10'-0"

Portable Metal

Optional Guard Rails 1" Pipe Welded
10'-0"
8'-0"
3'-0"

Cutting List

Item	No.	Description
A	12	2½" x 10'-0" pipe
B	16	2½" x 7" pipe
C	4	2½" x 8" pipe2
D	4	2½" x 6'-10" pipe
E	4	2½" x 8'-0" pipe

Optional Guard Rail

F	2	1" x 4'-0" pipe
G	1	1" x 2'-7" pipe
H	1	1" x 4'-8" pipe

2½" pipe, 5½" apart

5½"
12
6

Stiles and Passes

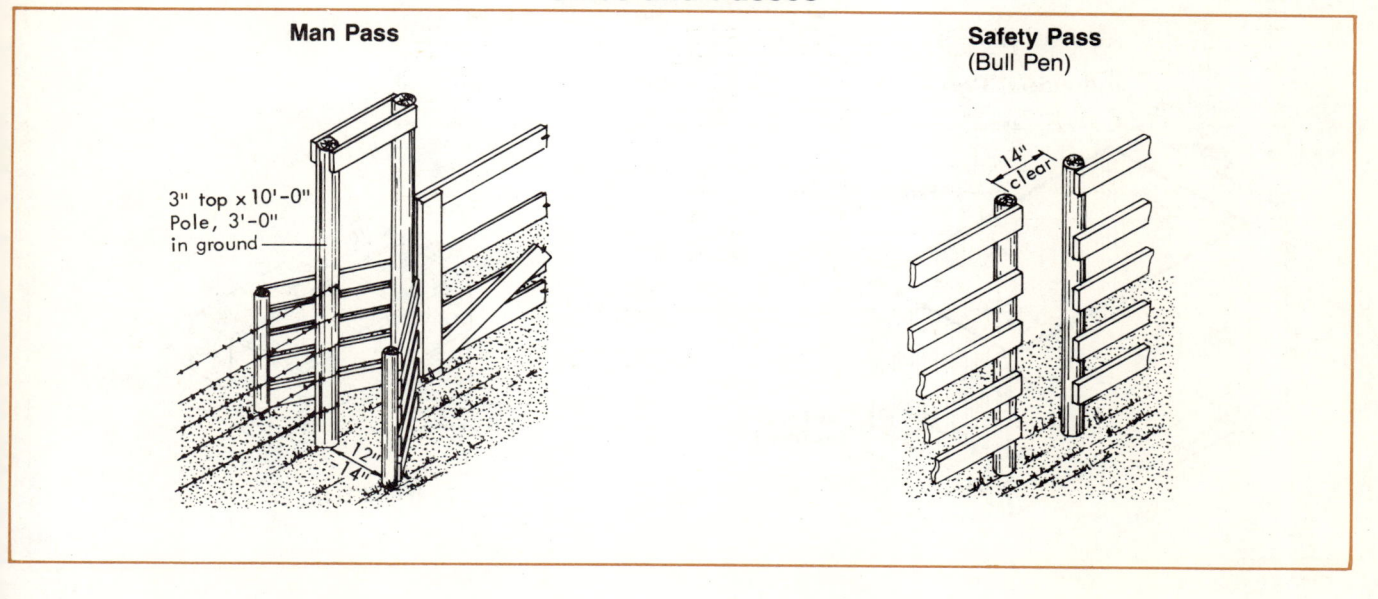

Man Pass

3" top x 10'-0" Pole, 3'-0" in ground

12"
14"

Safety Pass
(Bull Pen)

14" clear

Handling Equipment
Calf Stall and Pen

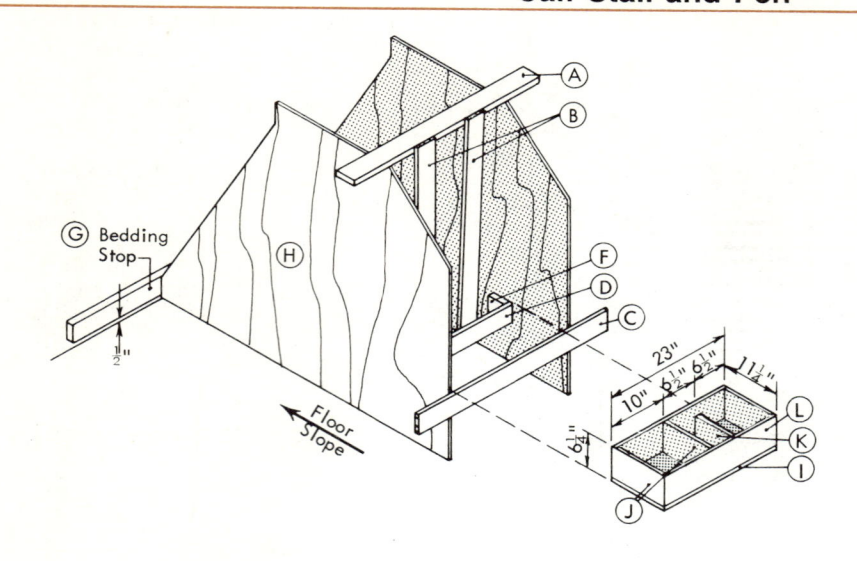

Cutting List

Item	No.	Description
A	1	2x4
B	2	1x4 x 3'-7"
C	1	1x4
D	1	1x4 x 23¼"
E	1	1x2 x 23¼"
F	2	2x4 block
G	1	2x4
H	1	⅜" plywood partition

Feedbox

Item	No.	Description
I	1	1x12 x 23"
J	3	1x6 x 9¾"
K	1	1x4 x 9¾"
L	2	1x6 x 23"

Floor Level Tie Stall

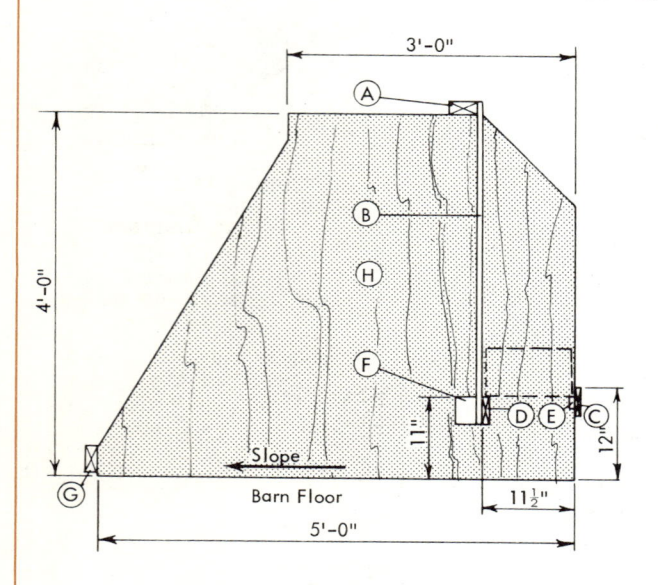

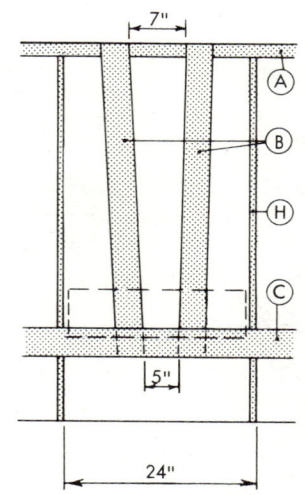

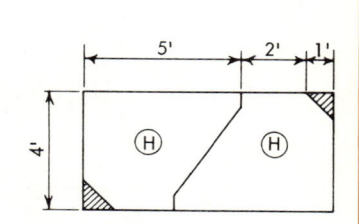

Plywood Cutting Diagram

Individual Pen

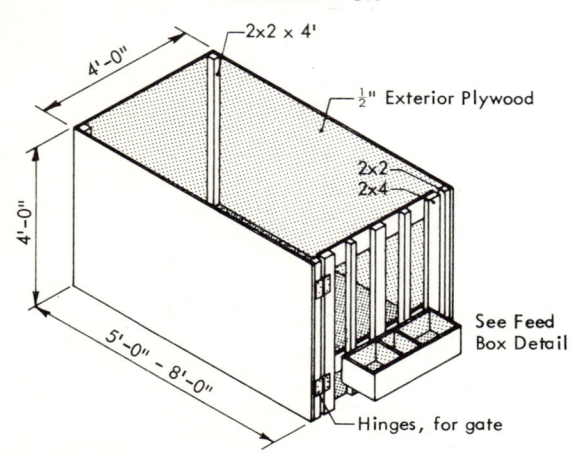

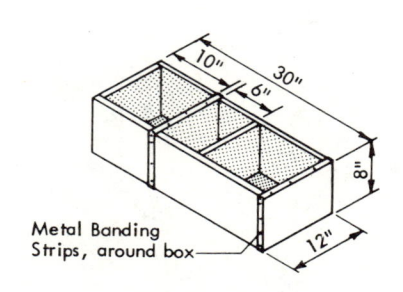

Feed Box

12.21

Movable Calf Hutch

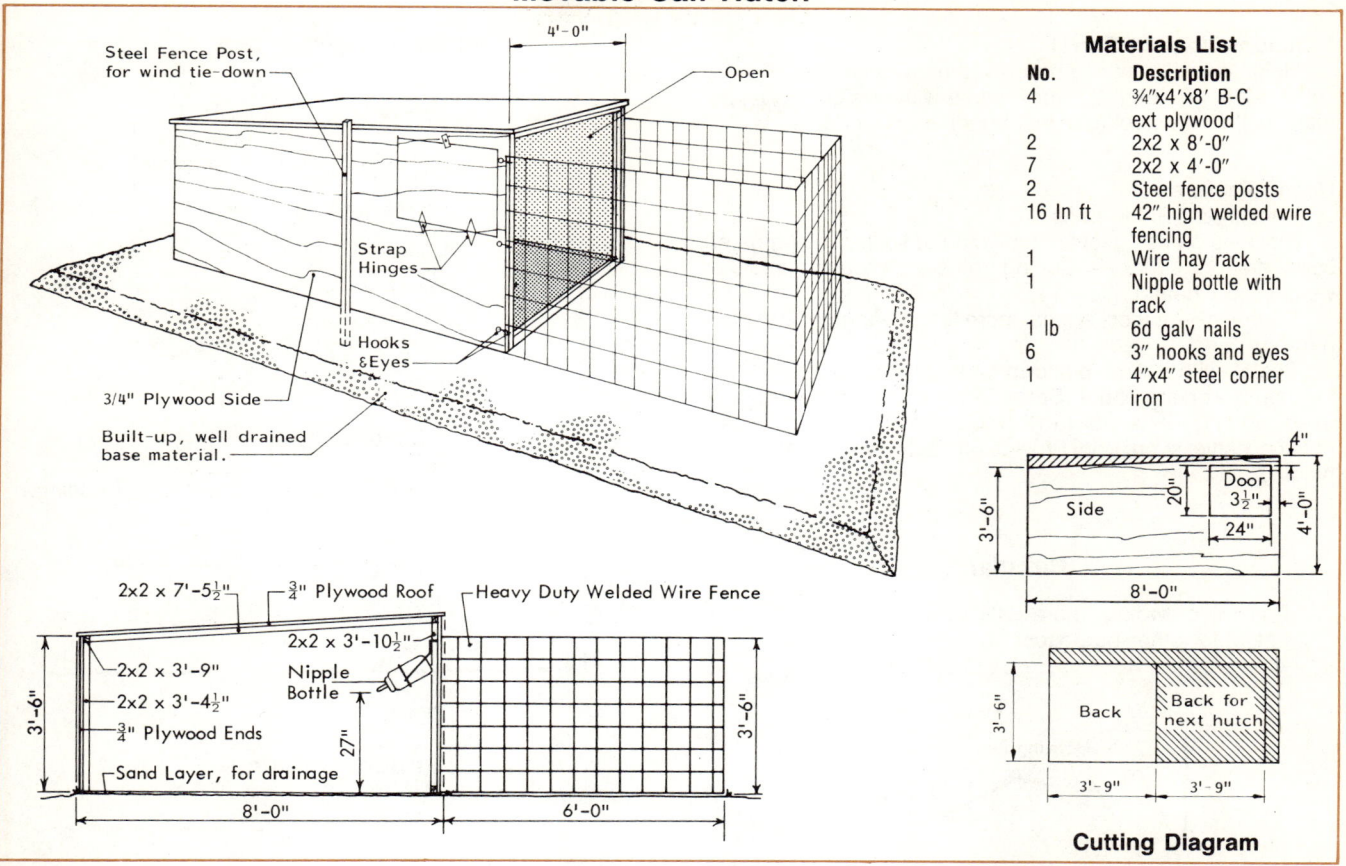

Materials List

No.	Description
4	¾"x4'x8' B-C ext plywood
2	2x2 x 8'-0"
7	2x2 x 4'-0"
2	Steel fence posts
16 ln ft	42" high welded wire fencing
1	Wire hay rack
1	Nipple bottle with rack
1 lb	6d galv nails
6	3" hooks and eyes
1	4"x4" steel corner iron

Cutting Diagram

Calf Hay Feeders

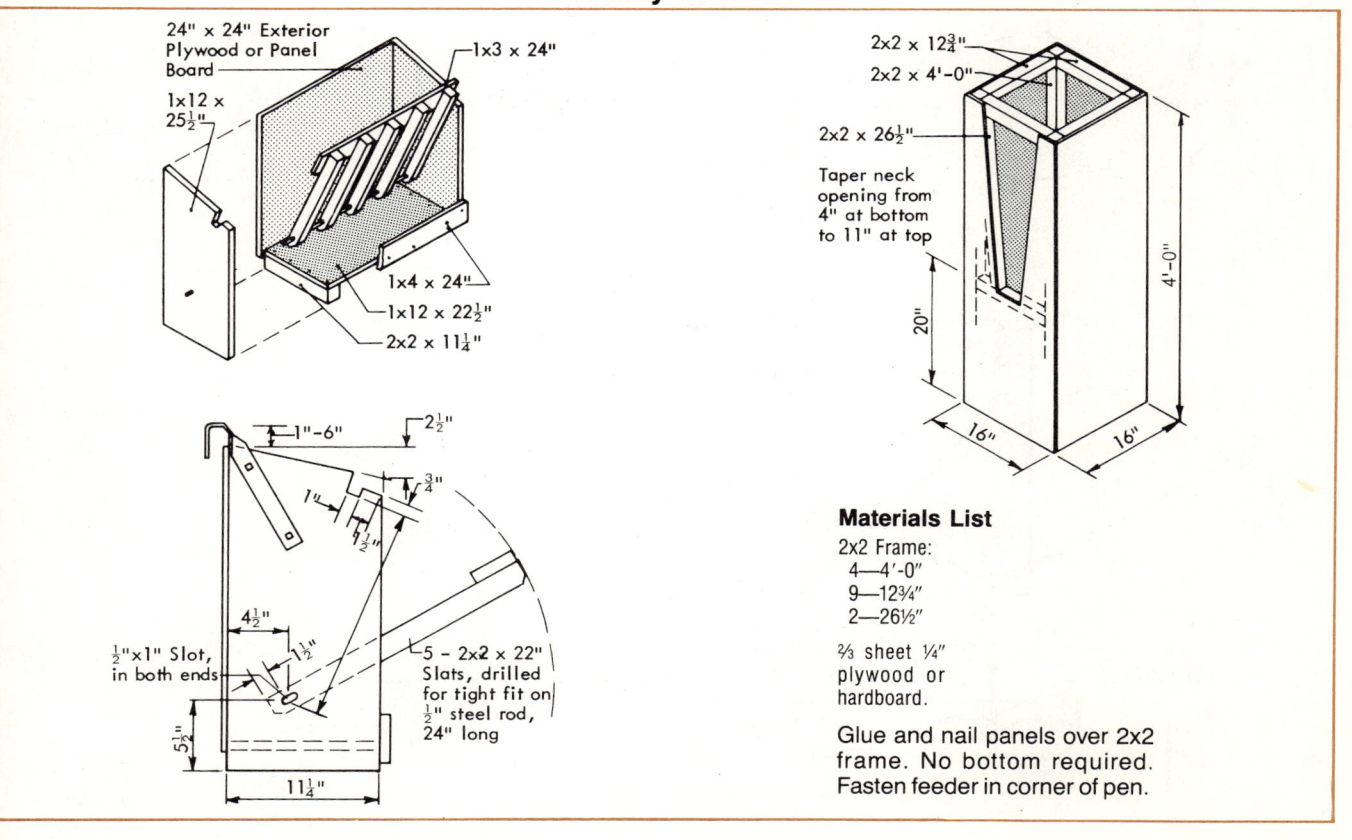

Materials List

2x2 Frame:
  4—4'-0"
  9—12¾"
  2—26½"

⅔ sheet ¼" plywood or hardboard.

Glue and nail panels over 2x2 frame. No bottom required. Fasten feeder in corner of pen.

Corrals

Minimum Working Corral

Holding and crowding pen, 15 ft²/mature animal. Working chute, 18 to 30′ long, 30″ max. width with straight sides; 34″ max. width with sloping sides. Loading chute, 30″ wide.

Useful Features

Additional holding pens.

Blocking gates—prevent crowding at scale, cutting gates, spray area, or squeeze. Cutting gates—separate animals by weight, age, health, etc.

Squeeze—restricts animal more than headgate, for veterinary services.

Scale—either large platform for truck or stock or portable, for use in working chute. Spray Pen—crowding pen may be used. Sorting Alley—10′ to 12′ wide. Provide water in lots for holding cattle overnight. Provide for feed and water in one holding pen.

Funnel Crowding Pen

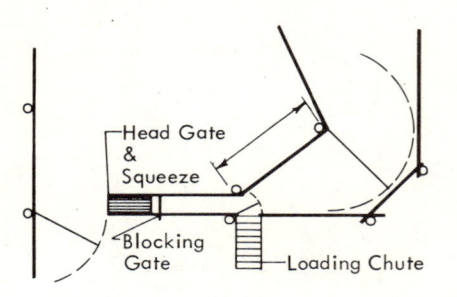

Sorting Alley

With cross gate, can be used for spraying and holding.

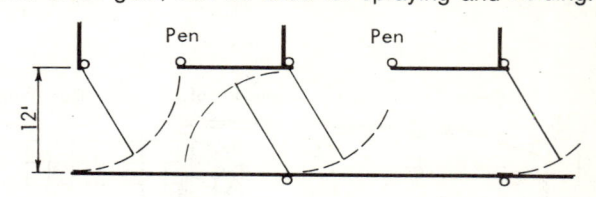

Circular

Positive crowding, more difficult to construct, less useful for other purposes.

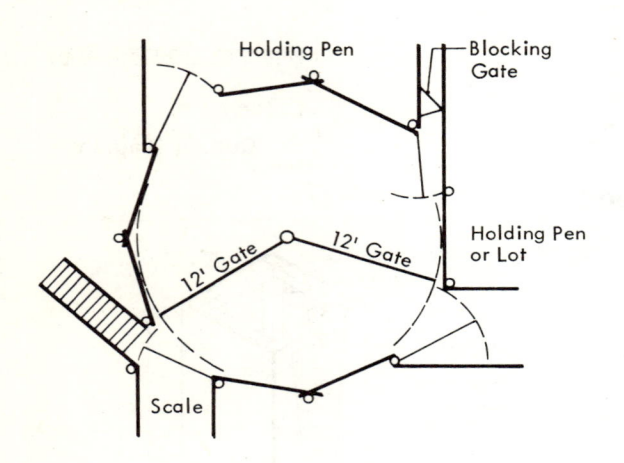

In Barn Corner

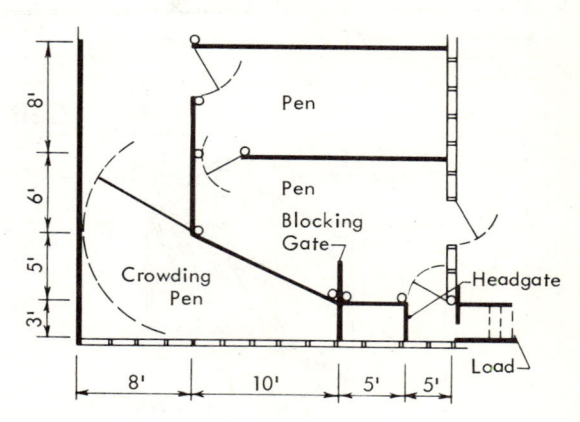

Small Square Corral

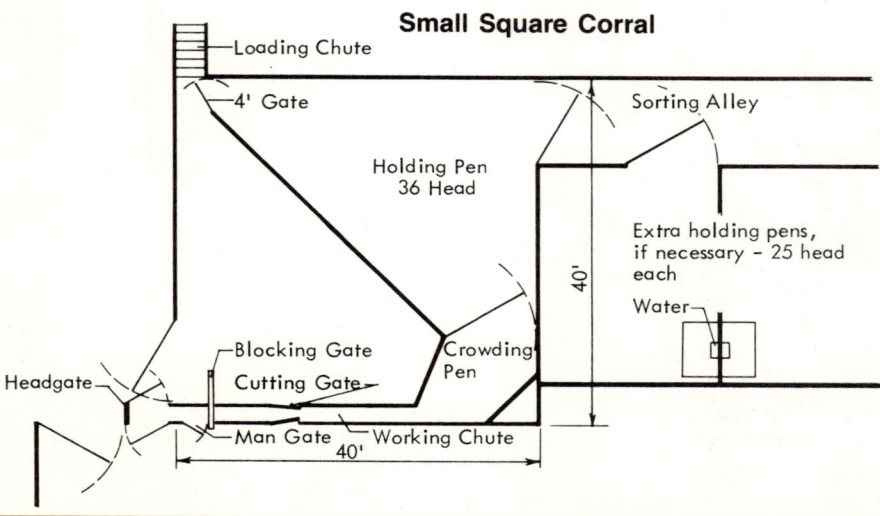

Loading Chutes

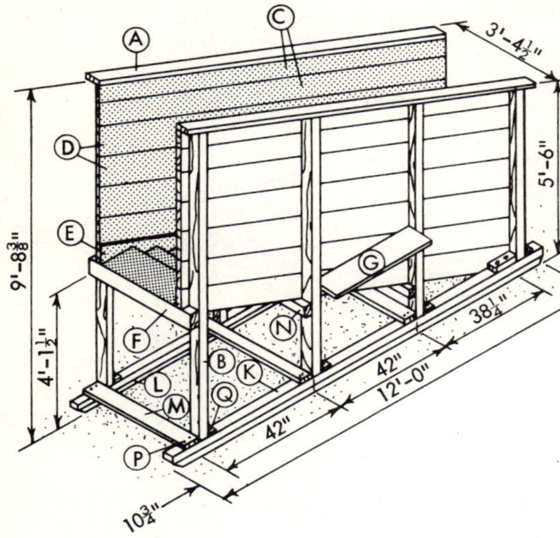

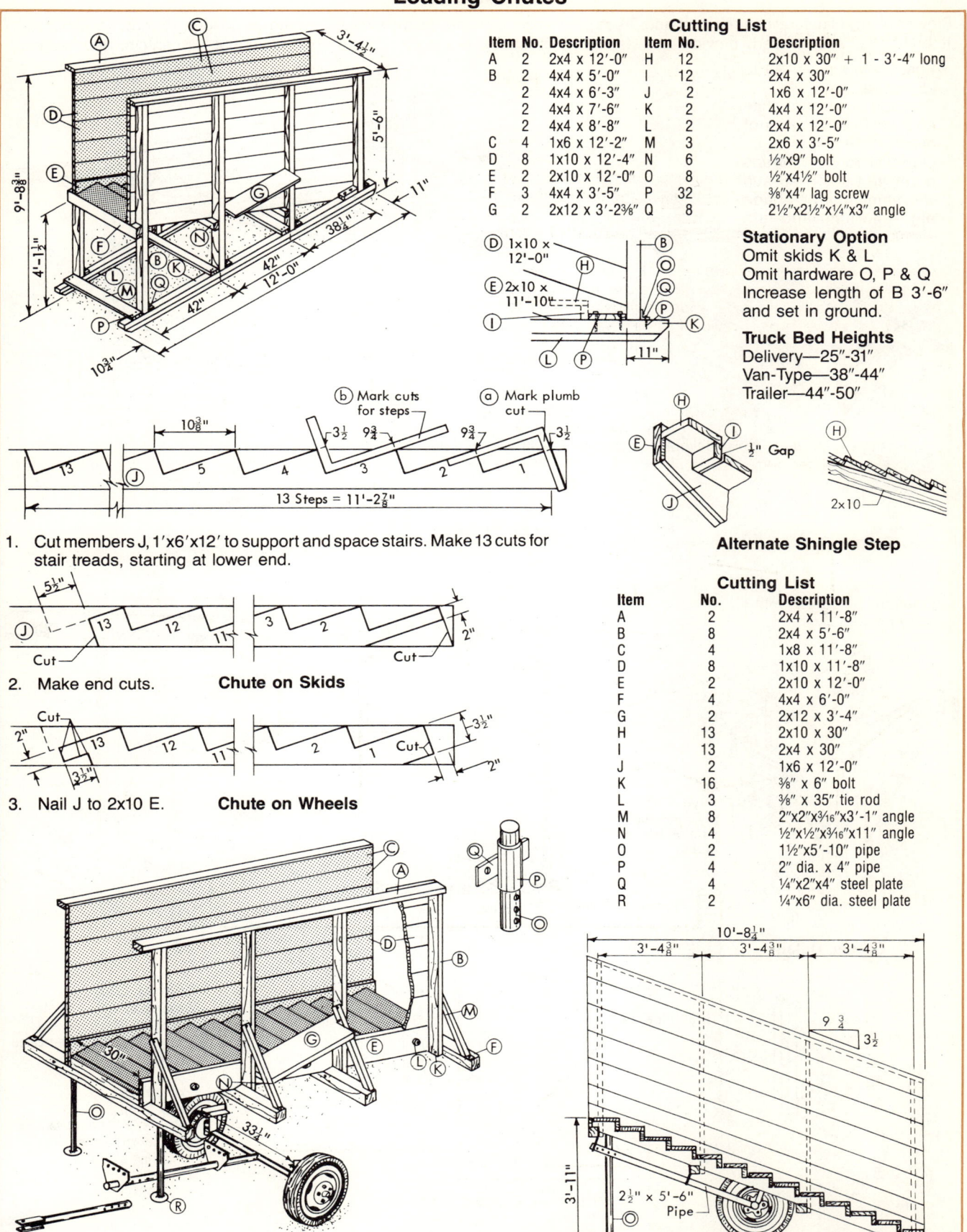

Cutting List

Item	No.	Description	Item	No.	Description
A	2	2x4 x 12'-0"	H	12	2x10 x 30" + 1 - 3'-4" long
B	2	4x4 x 5'-0"	I	12	2x4 x 30"
	2	4x4 x 6'-3"	J	2	1x6 x 12'-0"
	2	4x4 x 7'-6"	K	2	4x4 x 12'-0"
	2	4x4 x 8'-8"	L	2	2x4 x 12'-0"
C	4	1x6 x 12'-2"	M	3	2x6 x 3'-5"
D	8	1x10 x 12'-4"	N	6	½"x9" bolt
E	2	2x10 x 12'-0"	O	8	½"x4½" bolt
F	3	4x4 x 3'-5"	P	32	⅜"x4" lag screw
G	2	2x12 x 3'-2⅜"	Q	8	2½"x2½"x¼"x3" angle

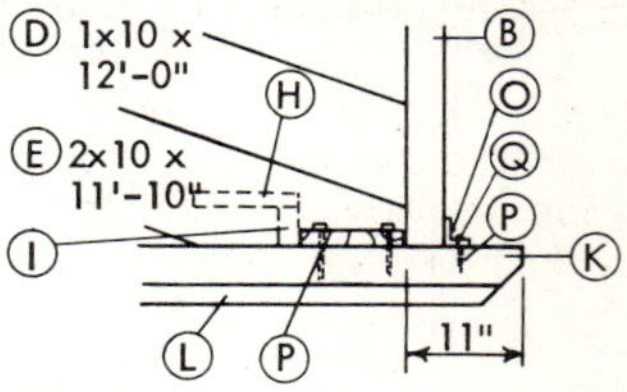

Stationary Option

Omit skids K & L

Omit hardware O, P & Q

Increase length of B 3'-6" and set in ground.

Truck Bed Heights

Delivery—25"-31"

Van-Type—38"-44"

Trailer—44"-50"

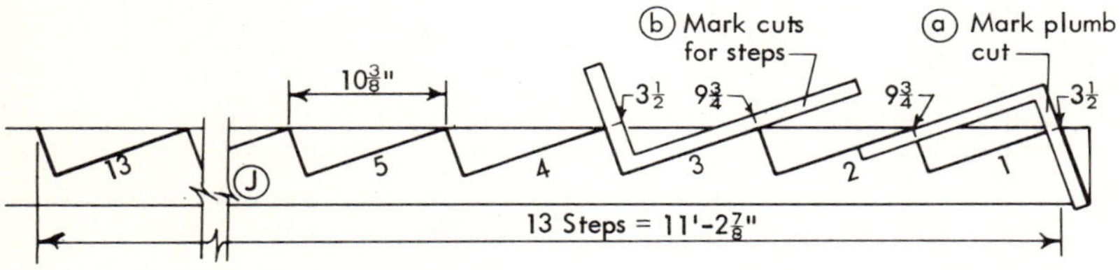

1. Cut members J, 1'x6'x12' to support and space stairs. Make 13 cuts for stair treads, starting at lower end.

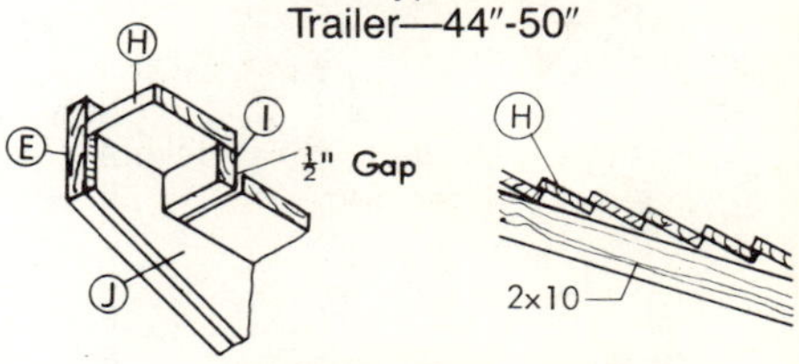

Alternate Shingle Step

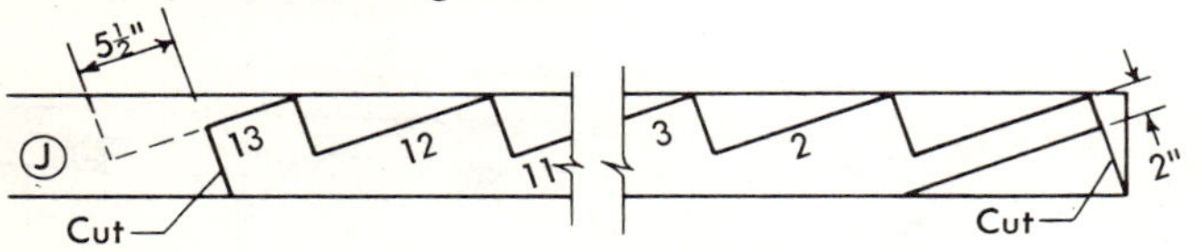

2. Make end cuts. **Chute on Skids**

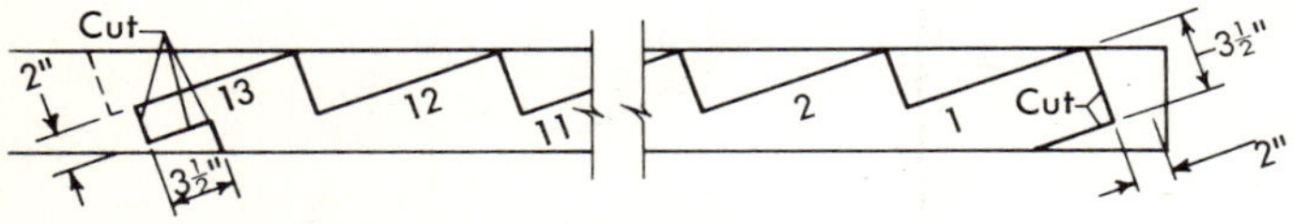

3. Nail J to 2x10 E. **Chute on Wheels**

Cutting List

Item	No.	Description
A	2	2x4 x 11'-8"
B	8	2x4 x 5'-6"
C	4	1x8 x 11'-8"
D	8	1x10 x 11'-8"
E	2	2x10 x 12'-0"
F	4	4x4 x 6'-0"
G	2	2x12 x 3'-4"
H	13	2x10 x 30"
I	13	2x4 x 30"
J	2	1x6 x 12'-0"
K	16	⅜" x 6" bolt
L	3	⅜" x 35" tie rod
M	8	2"x2"x³⁄₁₆"x3'-1" angle
N	4	½"x½"x³⁄₁₆"x11" angle
O	2	1½"x5'-10" pipe
P	4	2" dia. x 4" pipe
Q	4	¼"x2"x4" steel plate
R	2	¼"x6" dia. steel plate

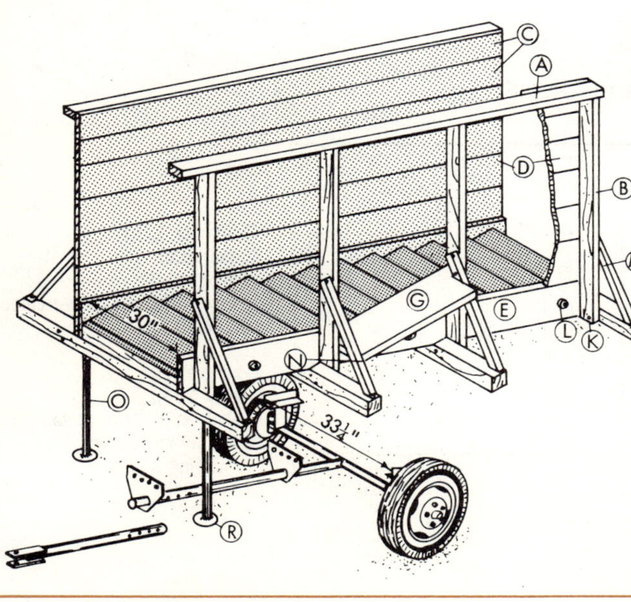

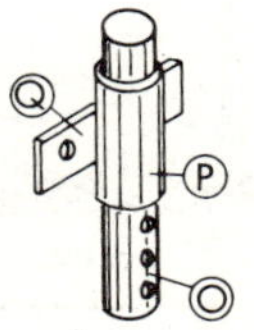

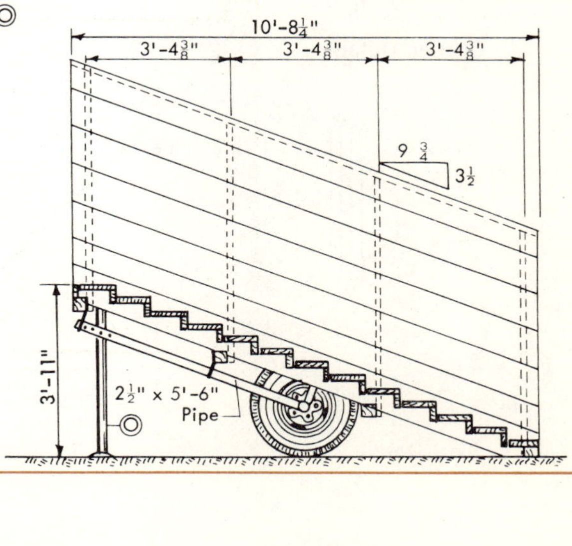

Sunshades

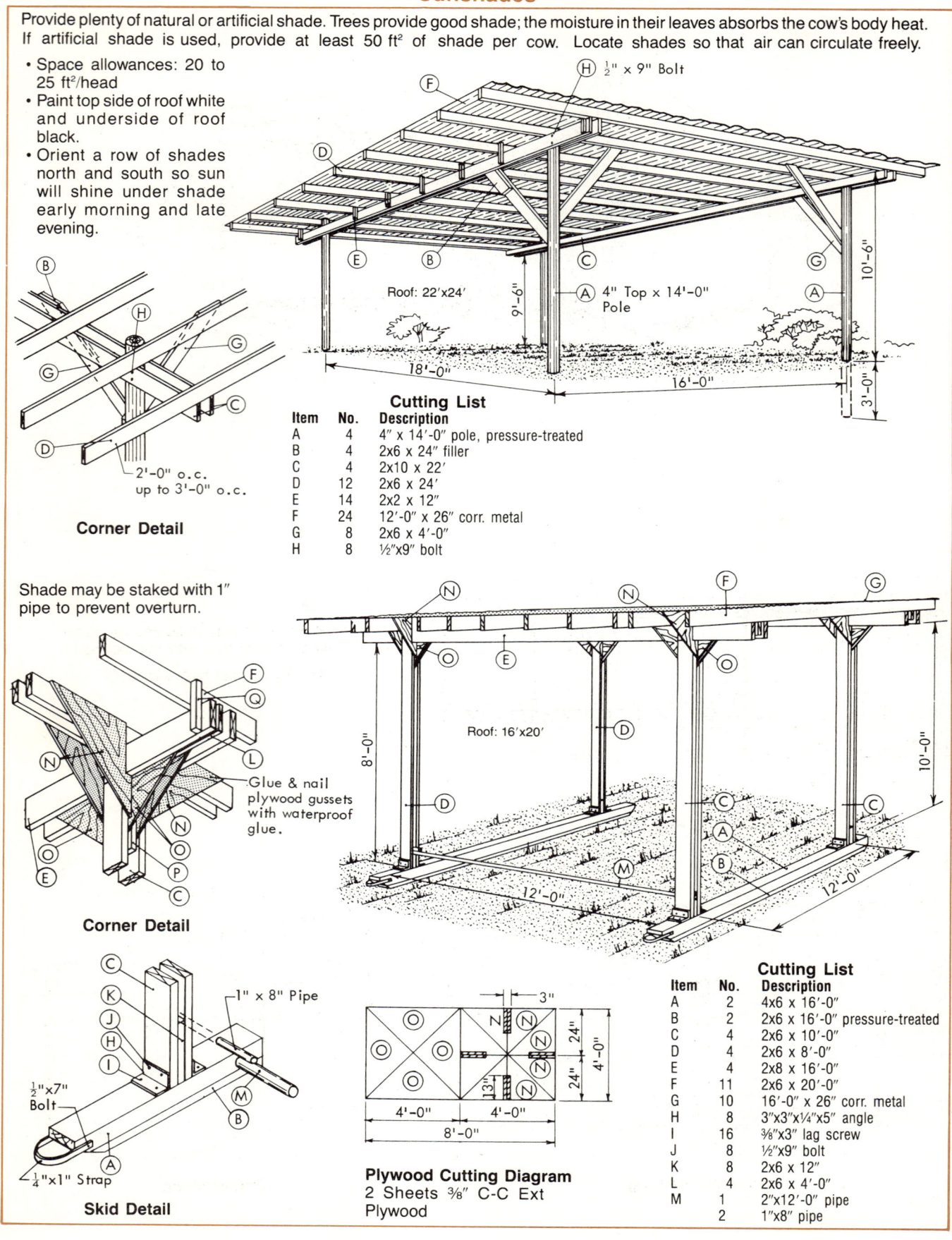

Provide plenty of natural or artificial shade. Trees provide good shade; the moisture in their leaves absorbs the cow's body heat. If artificial shade is used, provide at least 50 ft² of shade per cow. Locate shades so that air can circulate freely.

- Space allowances: 20 to 25 ft²/head
- Paint top side of roof white and underside of roof black.
- Orient a row of shades north and south so sun will shine under shade early morning and late evening.

Corner Detail

Cutting List

Item	No.	Description
A	4	4" x 14'-0" pole, pressure-treated
B	4	2x6 x 24" filler
C	4	2x10 x 22'
D	12	2x6 x 24'
E	14	2x2 x 12"
F	24	12'-0" x 26" corr. metal
G	8	2x6 x 4'-0"
H	8	½"x9" bolt

Shade may be staked with 1" pipe to prevent overturn.

Glue & nail plywood gussets with waterproof glue.

Corner Detail

Skid Detail

Plywood Cutting Diagram
2 Sheets ⅜" C-C Ext Plywood

Cutting List

Item	No.	Description
A	2	4x6 x 16'-0"
B	2	2x6 x 16'-0" pressure-treated
C	4	2x6 x 10'-0"
D	4	2x6 x 8'-0"
E	4	2x8 x 16'-0"
F	11	2x6 x 20'-0"
G	10	16'-0" x 26" corr. metal
H	8	3"x3"x¼"x5" angle
I	16	⅜"x3" lag screw
J	8	½"x9" bolt
K	8	2x6 x 12"
L	4	2x6 x 4'-0"
M	1	2"x12'-0" pipe
	2	1"x8" pipe

12.25

Calf Shelter

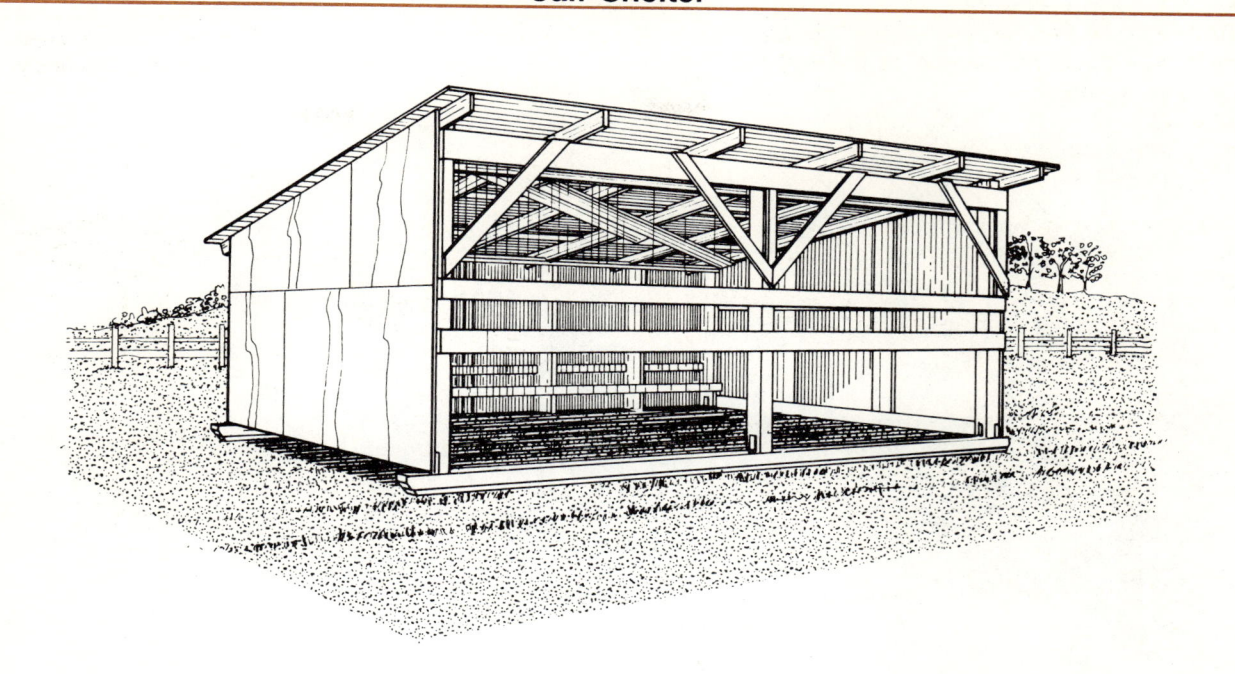

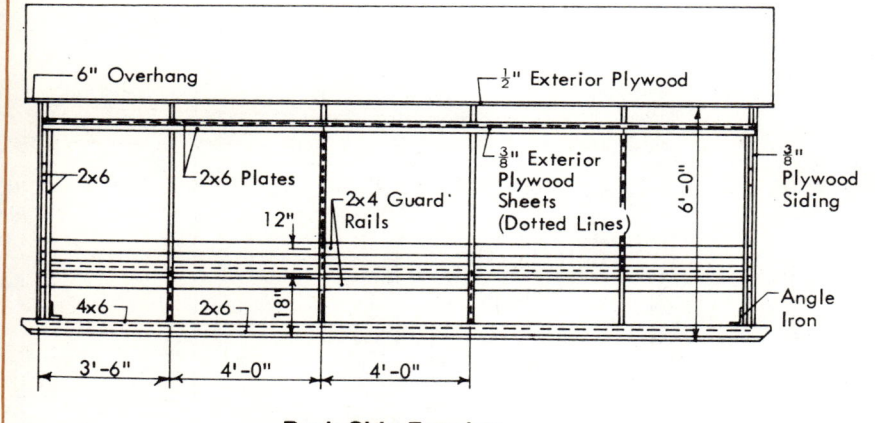

Back Side Framing

- 6" Overhang
- ½" Exterior Plywood
- 2x6
- 2x6 Plates
- 2x4 Guard Rails
- ⅜" Exterior Plywood Sheets (Dotted Lines)
- ⅜" Plywood Siding
- 6'-0"
- 12"
- 4x6
- 2x6
- 18"
- Angle Iron
- 3'-6", 4'-0", 4'-0"

Note:
Plywood sheets indicated by dotted lines. Cover roof with roll roofing.
Anchor against high winds.

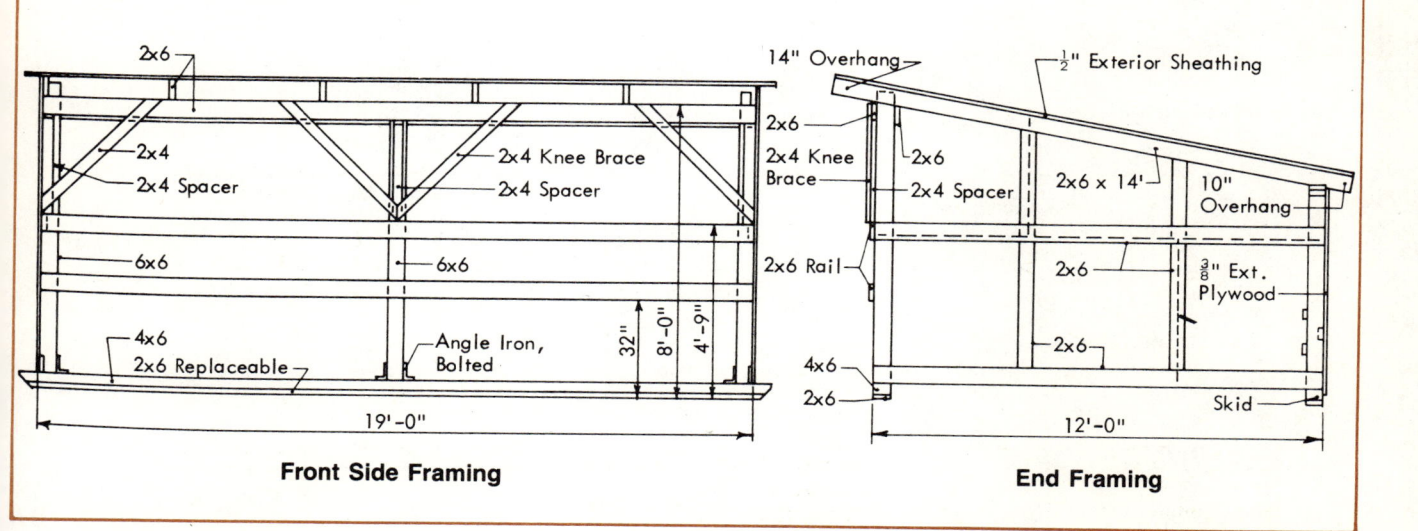

Front Side Framing

- 2x6
- 2x4
- 2x4 Spacer
- 2x4 Knee Brace
- 2x4 Spacer
- 6x6
- 6x6
- 4x6
- 2x6 Replaceable
- Angle Iron, Bolted
- 32"
- 8'-0"
- 4'-9"
- 19'-0"

End Framing

- 14" Overhang
- ½" Exterior Sheathing
- 2x6
- 2x4 Knee Brace
- 2x6
- 2x4 Spacer
- 2x6 x 14'
- 10" Overhang
- 2x6 Rail
- 2x6
- ⅜" Ext. Plywood
- 2x6
- 4x6
- 2x6
- Skid
- 12'-0"

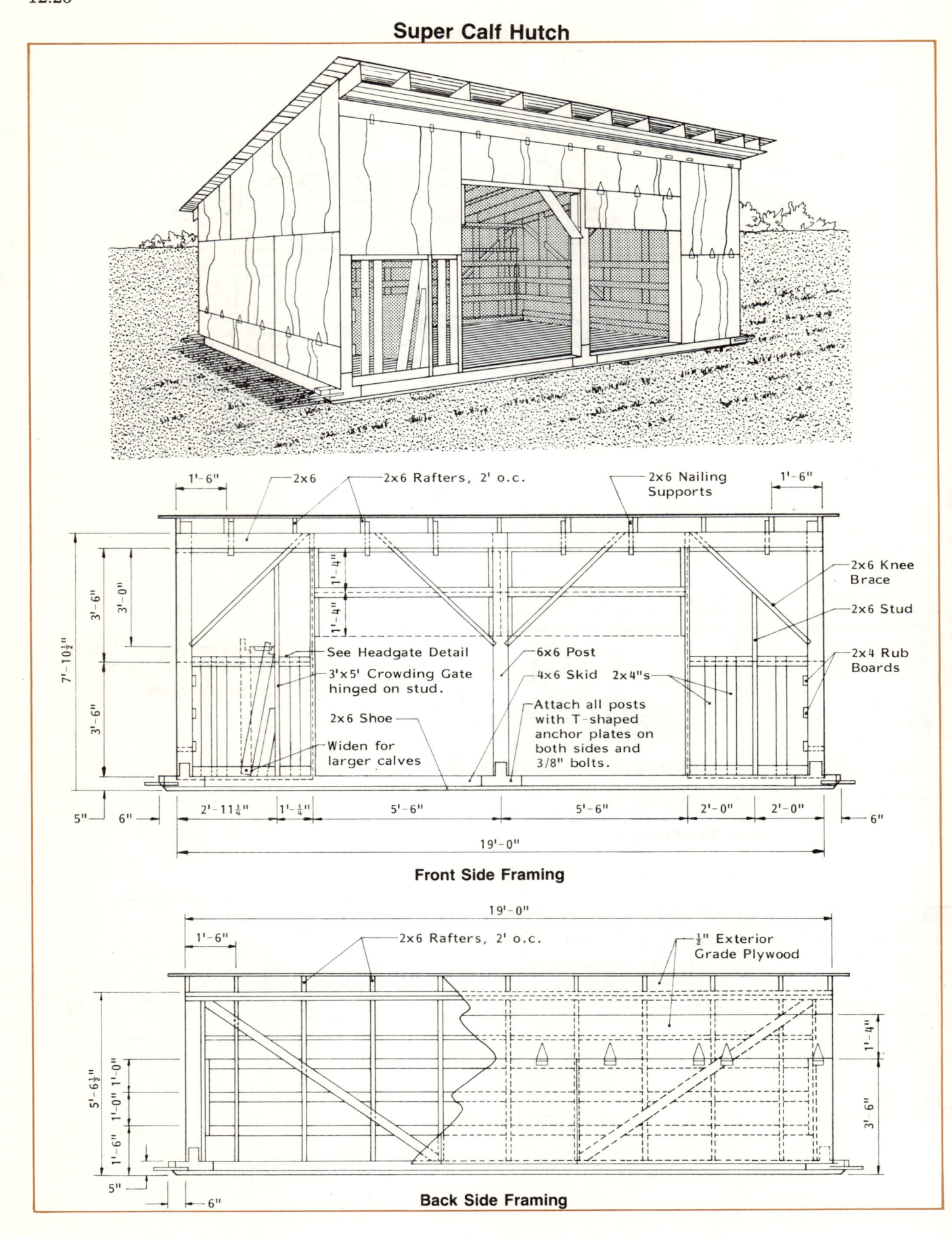

Super Calf Hutch

Front Side Framing

Back Side Framing

Super Calf Hutch, continued

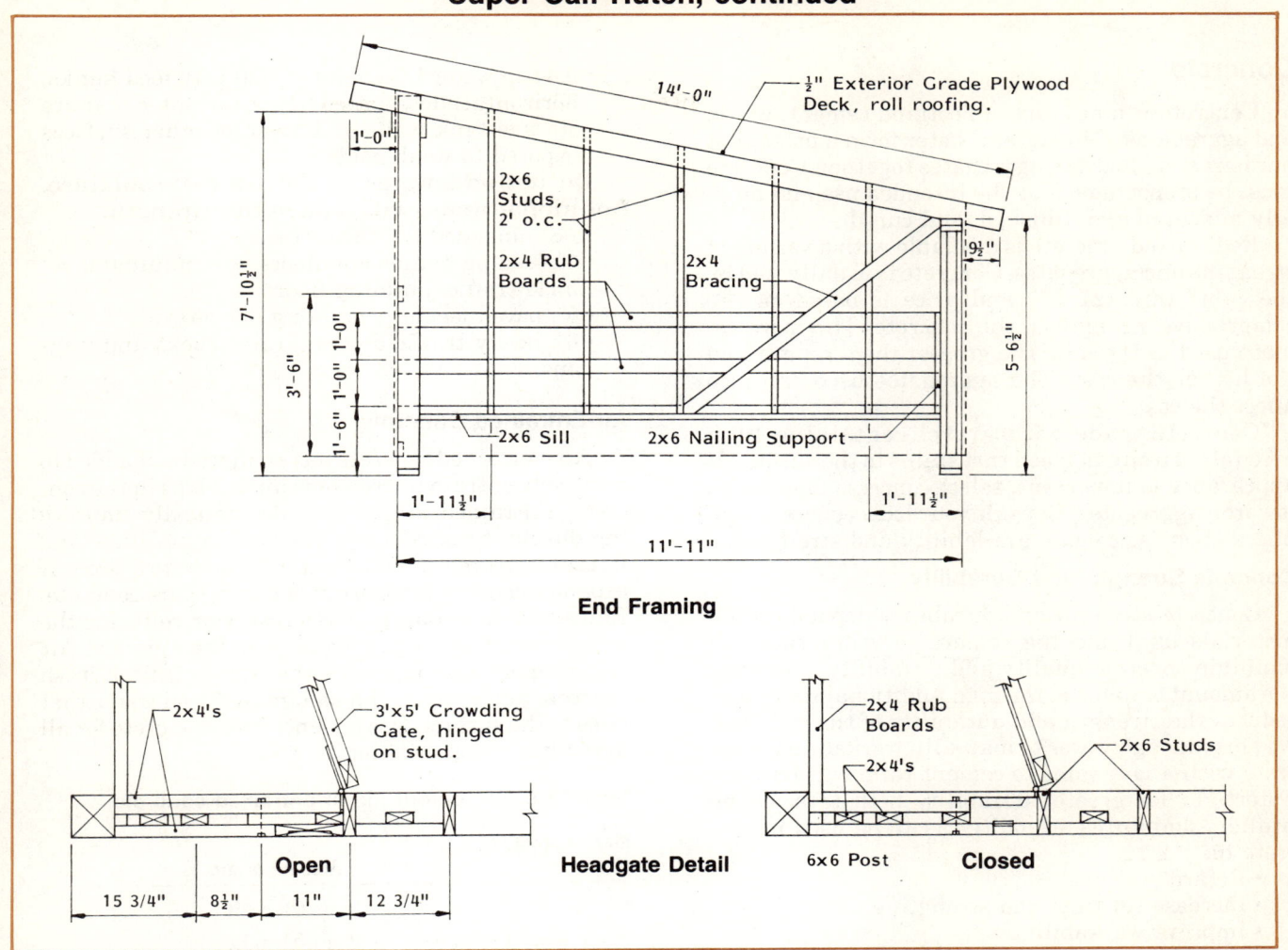

End Framing

- ½" Exterior Grade Plywood Deck, roll roofing.
- 2x6 Studs, 2' o.c.
- 2x4 Rub Boards
- 2x4 Bracing
- 2x6 Sill
- 2x6 Nailing Support

Dimensions: 14'-0", 1'-0", 7'-10½", 9½", 5'-6½", 3'-6", 1'-6", 1'-0", 1'-0", 1'-11½", 1'-11½", 11'-11"

Headgate Detail

- 2x4's
- 3'x5' Crowding Gate, hinged on stud.
- 2x4 Rub Boards
- 2x4's
- 2x6 Studs
- 6x6 Post

Open: 15 3/4", 8½", 11", 12 3/4"

Closed

Working Chute

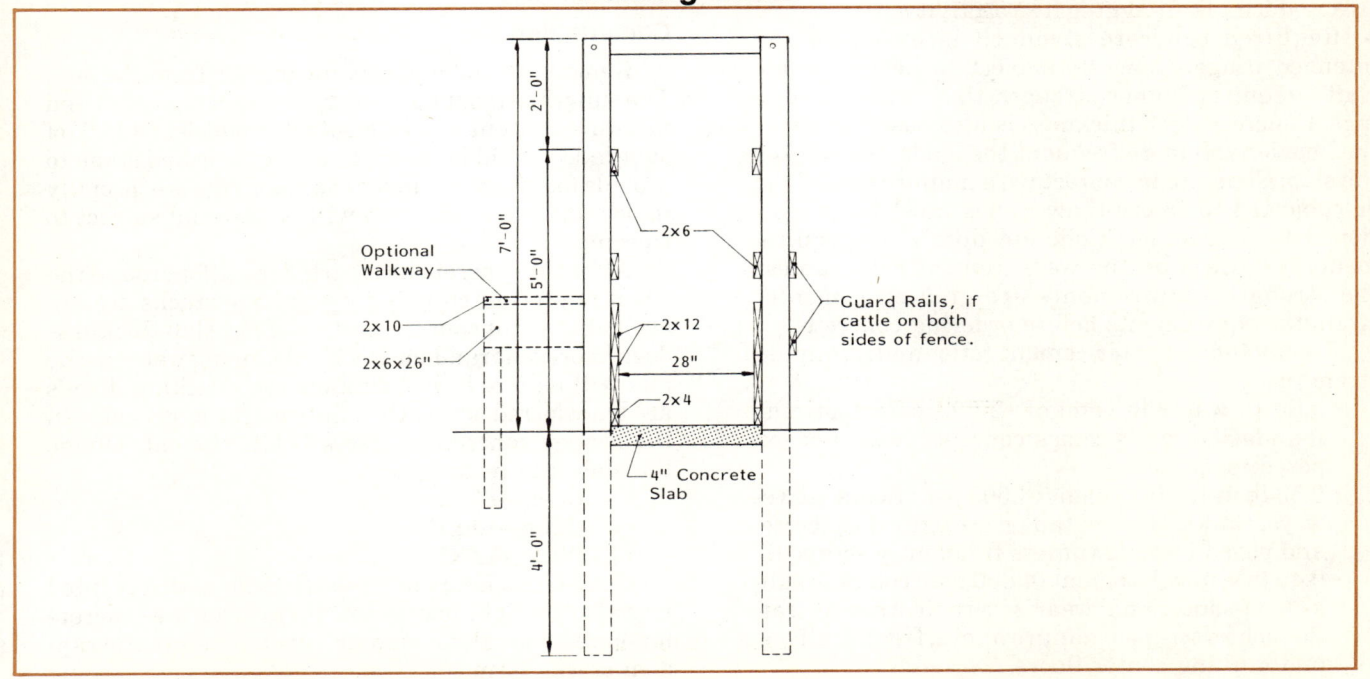

- Optional Walkway
- 2x10
- 2x6x26"
- 2x6
- 2x12
- 2x4
- 4" Concrete Slab
- Guard Rails, if cattle on both sides of fence.

Dimensions: 2'-0", 7'-0", 5'-0", 28", 4'-0"

12.27

13. CONCRETE

Concrete

Concrete is a mixture of Portland cement, water, and aggregates. Cement and water form a paste that hardens and glues the aggregates together. Concrete must be proportioned for the intended use and properly mixed, placed, finished, and cured.

Redi-mixed concrete is available with a variety of strengths and aggregates. Concrete is usually sold by the cubic yard (27 ft³), and price is based on the compressive strength of the concrete. The more cement used in the mix, the greater the strength and the higher the cost. The aggregates used can also affect the cost.

Concrete made with gravel containing high amounts of iron can bleed rust stains to the surface. If appearance is important, select concrete made with low iron aggregates. Consult your local concrete suppliers about aggregate availability and strengths.

Concrete Strength and Durability

Concrete strength and durability depend on the materials used and the cement to water ratio. To maintain concrete quality and durability, minimize the amount of water in the mix. Additional water also reduces the strength and durability of the concrete. Add just enough water to maintain workability. Concrete with a low water to cement ratio will be more watertight for greater durability both to acid and traffic. Admixtures or additives can be used in concrete to:
- Retard set time.
- Increase set time and strength gain.
- Improve workability.
- Entrain air.
- Increase strength.
- Reduce alkali-aggregate reactivity.

Required concrete strength is based on the intended usage. Concrete subject to heavy vehicle traffic requires higher strength than building footings. Concrete floor thickness is also based on use—the heavier and more frequent the loads, the thicker the slab. Concrete in contact with manure and silage or subjected to freeze-thaw cycles must be proportioned for durability. Concrete durability requirements are based on the water-cement ratio, unless the strength requirements are greater. Consider strength requirements before ordering concrete.

Recommended water-cement ratios and minimum strengths:
- 0.60 lb water/lb cement (2,500 psi): footings, foundation walls, mass concrete, etc., not exposed to weather.
- 0.53 lb water/lb cement (3,500 psi): floors, driveways, walks, storage tanks, structural concrete, and paved feedlots (unless frequently scraped).
- 0.49 lb water/lb cement (4,000 psi): concrete subjected to additional wear, severe weather, or traffic such as scraped manure alleys, feeding alleys, and milking center floors.
- 0.44 lb water/lb cement (4,500 psi): feed bunks, horizontal silos, slotted floors, unvented manure storage tank walls and cover, or other surfaces exposed to weak acids.

Do not add water to the concrete mixture. Additional water will reduce the strength.

Recommended floor thickness:
- 4″: feeding aprons and floors with minimum vehicle traffic, building floors.
- 5″: paved feedlots, building driveways.
- 6″: heavy traffic drives (grain trucks and wagons).

Air-Entrained Concrete

Air-entrained concrete has an ingredient added to purposely entrain microscopic air bubbles in the concrete. Entrained air bubbles dramatically improve the durability of concrete exposed to moisture and freeze-thaw cycles. As the water in concrete freezes, it expands, causing pressure that can rupture concrete. Entrained air bubbles act as a reservoir, relieving the pressure and preventing damage to the concrete. Air entrainment also improves the workability of fresh concrete and reduces the amount of bleed water that rises to the surface. Use air-entrained concrete for all agricultural construction.

Table 13-1. Air content for air-entrained concrete.

Max. aggregate size	Amount of air, %
1½″, 2″, or 2½″	5% ± 1%
¾″ or 1″	6% ± 1%
⅜″ or ½″	7½% ± 1%

Construction

Remove all sod and organic matter from the site. The subgrade must have uniform soil compaction and moisture content and be easily drained. The top 6″ of subgrade should be sand, gravel, or crushed stone to provide for drainage under the slab. This is especially important where the slab will be wet and subject to freezing.

After the concrete has started to stiffen, round the edges to prevent chipping. Shrinkage cracks are unavoidable so cut control joints ¼ of the slab thickness deep to prevent random cracks. Reducing water in the mix will help reduce shrinkage and cracking. Loads are transferred across the joints by the aggregates in the broken concrete surfaces below the cut. Divide the slab into rectangles:
- 4″ thick—8′x12′.
- 5″ thick—10′x15′.
- 6″ thick—12′x18′.

Cut control joints in fresh concrete with a pointed trowel or straight hoe, or saw them after the concrete has cured enough for smooth cuts but before the random cracks form.

Isolation and expansion joints permit the slab to move with earth and temperature changes. Place ½" wide isolation joints along existing improvements such as buildings, concrete water tanks, or paved drives. Install expansion joints in long walks and drives to prevent buckling of the slab during hot weather. Increase the slab edge thickness to help reduce cracking, resist equipment damage, and to allow for soil erosion.

Additional information on the mixing and placing of concrete is available in the Portland Cement Association publications *Design and Control of Concrete Mixtures, Cement Mason's Guide,* and *Concrete-Paved Feedlots.*

Curing

Concrete does not dry—it hardens by a chemical reaction called hydration. Excessive evaporation from the surface of fresh concrete reduces its ultimate strength. Cure fresh concrete by covering with a plastic film, continuously ponding with water, covering with wet burlap or straw, or applying a curing compound. Be sure to cure concrete walls by one of these methods after the forms are removed. Proper curing can increase concrete strength about 50% over concrete allowed to air dry after finishing. Begin curing as soon as the concrete surface is hard enough not to be damaged by the water. Continue this for 5 to 7 days.

Slip Resistant Concrete Floors

In livestock housing, wood float and broom finished surfaces become smooth in time due to tractor scraping and constant animal traffic. Select the degree of floor roughness based on the intended use and animal type.

New floors can be scored or grooved with a homemade tool as they are poured, Figs 13-1 and 13-2. Make grooves ⅜"-½" deep by ½"-1" wide, spaced 4"-8" apart. Make grooves diagonal to the direction of animal traffic. Deep grooves make cleaning and disinfecting more difficult and can cause foot and leg problems in smaller animals.

With new floors, where disinfection is required and grooves present a problem, aluminum oxide can be added to the surface when the floor is poured. Apply aluminum oxide grit (as in sandpaper) at ¼ to ½ lb/ft^2 before the concrete sets. Coarse grit (4 to 6 meshes/in) is recommended.

Existing slick floors can be grooved with a mechanical grinder (similar to ones used for sawing concrete) or painted with chlorinated rubber paint or an exterior latex. For additional traction, sprinkle with sawdust or coarse ground cornmeal while paint is wet. Repaint annually.

Badly worn but sound floors can be resurfaced with a concrete overlay. Bonding a thin overlay to an existing floor requires special cleaning and concrete mixes. Refer to the Portland Cement Association publication *Resurfacing Concrete Floors* for additional information on floor resurfacing.

For more information about slip resistant floors, refer to Midwest Plan Service's AED-19, *Slip Resistant Concrete Floors.*

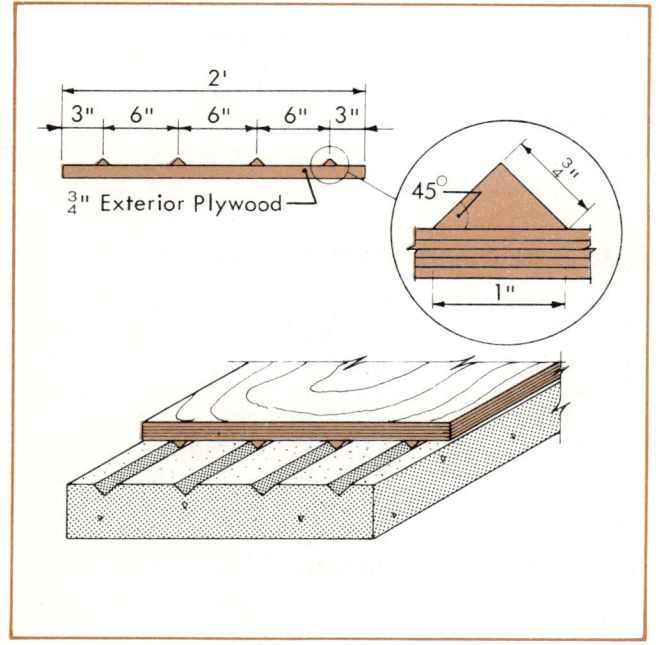

Fig 13-1. Wood groover.

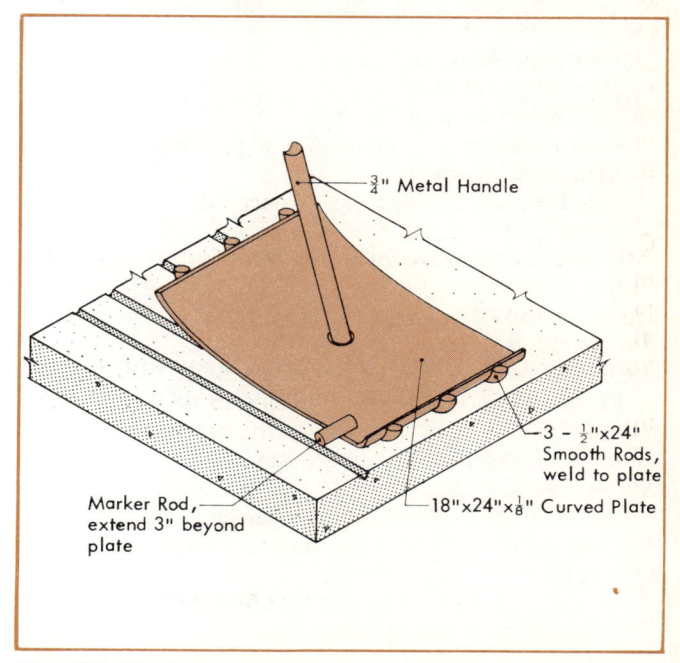

Fig 13-2. Steel groover.

14. SELECTED REFERENCES

Available from the Extension Agricultural Engineer at any of the institutions listed on the inside front cover or from the Midwest Plan Service.

MWPS-1 *Structures and Environment Handbook.*

MWPS-2 *Farmstead Planning Handbook.*

MWPS-13 *Planning Grain-Feed Handling for Livestock and Cash-Grain Farms.*

MWPS-14 *Private Water Systems Handbook.*

MWPS-18 *Livestock Waste Facilities Handbook.*

AED-15 *Tilt-Up Concrete Horizontal Silo Construction.*

AED-18 *Selecting Dairy Manure Handling Systems.*

AED-19 *Slip Resistant Concrete Floors.*

AED-23 *Outside Liquid Manure Storages.*

TR-3 *Concrete Manure Tank Design.*

TR-4 *Welded Wire Fabric in Concrete Manure Tanks.*

TR-9 *Circular Concrete Manure Tanks.*

Available from Northeast Regional Agricultural Engineering Service (NRAES), Riley Robb Hall, Cornell University, Ithaca, NY 14853. Ph. 607/256-7654.

NRAES-1 *Pole and Post Buildings.*

NRAES-11 *High-Tensile Wire Fencing.*

NRAES-12 *Milking Center Design Manual.*

NRAES-15 *Planning Dairy Stall Barns.*

NRAES-18 *Extinguishing Silo Fires.*

FS-18 *Dairy Farm Heat Exchangers for Heating Water.*

FS-19 *In-Line Milk Cooling on the Farm.*

FS-33 *Troubleshooting Dairy Ventilating Systems.*

Other resources.

Agricultural Wiring Handbook. Food and Energy Council, Inc., Columbia, MO 65202. 1978.

Cement Mason's Guide. Portland Cement Association, Skokie, IL 60077-4321. 1980.

Chore Reduction for Confinement Stall Dairy Systems. Hoard's Dairyman, Fort Atkinson, WI 53538. 1978.

Chore Reduction for Free Stall Dairy Systems. Hoard's Dairyman, Fort Atkinson, WI 53538. 1978.

Concrete-Paved Feedlots. Portland Cement Association, Minneapolis, MN 55435.

Dairy Housing II, Proceedings of the Second National Dairy Housing Conference. American Society of Agricultural Engineers, St. Joseph, MI 49085. 1983.

Design and Control of Concrete Mixtures. Portland Cement Association, Skokie, IL 60077-4321. 1979.

Electrical Wiring Systems for Livestock and Poultry Facilities. H. David Currence, National Food and Energy Council, Inc., Columbia, MO 65202. 1983.

Farm Lighting Design Guide, SP-0175. American Society Agricultural Engineers, St. Joseph, MI 49085.

Fire In Silos, Prevention and Extinguishing. International Silo Association, Inc., West Des Moines, IA 50265.

Livestock Environment, Proceedings of the International Livestock Environment Symposium. American Society Agricultural Engineers, St. Joseph, MI 49085. 1974.

Livestock Environment II, Proceedings of the second international livestock environment symposium. American Society Agricultural Engineers, St. Joseph, MI 49085. 1982.

The National Electrical Code Handbook. National Fire Protection Association, Quincy, MA. 1980.

Resurfacing Concrete Floors. Portland Cement Association, Skokie, IL 60077-4321. 1981.

Stray Voltage Problems with Dairy Cows. R.D. Appleman and H.A. Cloud, Agricultural Extension Service, University of Minnesota, St. Paul, MN 55108. 1980.

15. INDEX

This is an index to key words based on titles and headings. See detailed table of contents, page i. See planning data summary, page 1.1.

Barn
    Free stall 4.4-4.8, 11.2
    Plan descriptions 4.4, 4.7, 4.8
    Remodeling 2.2, 4.4
    Stanchion or tie stall 4.1-4.4, 11.2
Bedding 8.2
Bulk tanks 5.4
Bunks
    Covered 12.3
    Fenceline 12.1, 12.2
    Lot 12.4

Calf
    Hay feeders 12.21
    Housing 3.2, 3.3
    Hutch 12.21
    Shelter 12.25, 12.26
    Stall and pen 12.20
Complete mixed rations 9.1
Concrete 13.1, 13.2
Conversion factors 1.2
Corrals 12.22

Drains 5.6

Electrical 11.1-11.3
Environment 7.1-7.15
Equipment plans 12.1-12.27

Fans 7.8-7.10
Farmstead planning 2.1-2.3
Feed centers 9.3
Feeders
    Hay 12.5-12.7, 12.21
    Mineral 12.9
    Supplemental concentrate 9.1
Feeding
    Dry hay 9.3
    Equipment 12.5-12.9
    Facilities 9.1-9.4
    Fences 12.5, 12.6
    Loose housing 9.1
    Space 1.1, 3.3
    Stall barn 9.1
Fences
    Braces and corners 12.12
    Electric 11.4-11.7
    Feeding 12.5, 12.6
    Stock guards 12.18, 12.19
    Windbreak 12.10, 12.11
Fencing 12.12-12.14

Floor
    Bedded pack 3.1
    Counter-slope 3.1
    Milking center 5.6
    Slip resistant concrete 13.2
    Slotted 3.1, 8.2
Flushing systems 8.3
Free stalls 3.1, 4.4-4.8

Gates 12.14-12.16
Gravity flow channel 8.3, 8.4
Gutters 4.3

Handling equipment 12.20-12.27
Heaters and heat exchangers 7.11
Herd makeup 2.1
Hinges 12.17
Holding area 5.2, 5.9
Holding pond 8.11, 8.12
Hospital area 5.10, 6.1, 6.2
Housing
    Calf 3.2, 3.3
    Dry cows 3.4
    Heifer 3.3
    Replacement animal 3.1-3.5
    Space 1.1, 3.1

Insulation 5.6, 7.11-7.15
Irrigation 8.10

Latches 12.17
Lighting 5.7, 11.2
Lightning 11.4
Loading chutes 6.2, 12.23
Lot runoff 8.10

Manure management 8.1-8.15
Maternity area 6.2
Milking center 5.1-5.11
    Construction 5.6-5.9
    Effluent 8.1
    Environment 5.9-5.11
Milking herd facilities 4.1-4.8
Milking parlors 5.1, 5.2, 5.10, 11.2
Milk room 5.4, 5.11, 11.2
Mineral feeders 12.9

Office 5.4, 5.11

Planning data 1.1, 1.2
Pumps 8.12-8.15

References 14.1

Septic tank 8.1
Settling basin 8.10, 8.11
Settling tank 8.1
Shelter 12.25, 12.26
Silage cart 12.8
Silo capacities 10.1-10.4
Space requirements
    Feeding 1.1, 3.3
    Housing 1.1, 3.1
Stall
    Arrangement 4.3
    Barns 4.1-4.4
    Calf 12.20
    Free 3.1, 4.4-4.8
    Mats 4.2
    Size 4.2
    Stanchion 4.1, 6.1
    Tie 4.1, 4.2
Stiles and passes 12.19
Stock guards 12.18, 12.19
Storage
    Equipment and supply room 5.4, 5.11
    Feed 10.3, 10.4
    Manure 8.4-8.8
    Silage 9.3, 10.1-10.4
Stray voltage 5.8
Sunshades 12.24

Treatment
    Facilities 6.1, 6.2
    Hospital area 5.10
    Stanchions 6.1, 6.2
Utilities 11.1-11.7
Utility room 5.4, 5.11

Vapor barriers 7.13
Vegetative infiltration area 8.12
Ventilating system maintenance 7.10
Ventilation
    Air inlets 7.7, 7.8
    Attic 7.11
    Eave, ridge, and wall openings 7.2, 7.3
    Exhaust 7.5, 7.7-7.9
    Manure storage 7.10
    Mechanical 7.4-7.11
    Natural 7.1-7.4

Water supply 5.5, 11.3-11.5
Windbreaks 12.10, 12.11
Working chute 12.27